Reliable Design of Electronic Equipment

Dhanasekharan Natarajan

Reliable Design of Electronic Equipment

An Engineering Guide

 Springer

Dhanasekharan Natarajan
Bangalore
India

ISBN 978-3-319-36168-0   ISBN 978-3-319-09111-2   (eBook)
DOI 10.1007/978-3-319-09111-2

Springer Cham Heidelberg New York Dordrecht London

Printed on acid-free paper

Springer is part of Springer Science+Business Media (www.springer.com)

To Om Shakthi

Preface

In the present competitive environment, organizations subcontract items such as power supply units, RF filters, attenuators, and RF amplifiers for delivering cost-effective electronic systems faster without compromising quality and reliability. Entrepreneurs venture into the design, development, and manufacturing of the electronic items. There are also entrepreneurs developing and manufacturing standard Commercially Off-The-Shelf (COTS) items such as DC–AC Inverters, Switched Mode Power Supply (SMPS), and LED lighting units. The COTS items offer second buying option for organizations.

This book presents the methods of applying reliability engineering techniques for the design and development of electronic equipment and COTS items. Reliability mathematics is limited to analyzing field failure data statistically. Adequate information for designing computerized reliability database system to support the application of the techniques is also presented. The presentation is based on Q&R literature, industrial experience in the design of RF Filters [1], and the application of the reliability techniques in the design phase of electronic equipment. The book provides excellent support for electrical and electronics engineering students for their two-credit elective reliability engineering course, bridging academic curriculum with industrial expectations.

Chapter 1 presents the overview of the methods of integrating reliability engineering techniques with the design and development activities of equipment. Chapter 2 explains the reliability related terms associated with the selection of electronic components. The terms are de-rating, ambient temperature, surface temperature, thermal resistance, and junction temperature. The stress factors, the quantum of de-rating, and information for ensuring the reliable application of electronic components are presented in Chap. 3.

Failure Mode and Effects Analysis (FMEA) focusses on the possible failures of equipment for improving reliability and maintainability. Applying FMEA is explained in Chap. 4 with examples from electronic designs benefitting circuit designers. FMEA terms, benefits of FMEA, software support, deriving maintenance flow diagrams, and library of preventive actions are also explained.

Proto-models of equipment are developed using the provisional engineering documents of design stage. Although it is a common practice to apply de-rating to components during design, the portion of failures due to overstresses is still very significant, approximately 25 % [2]. Overstress analysis is conducted on the equipment to ensure that stress applied on parts is within their ratings. The low-cost methods of conducting electrical, thermal, and vibration resonance overstress analyses are explained in Chap. 5 in tutorial form.

The proto-models of equipment are evaluated internally to verify compliance to contractual specifications prior to qualification testing. Reliability improvement tests are part of internal verification testing with the objective of precipitating design errors and achieving first-time acceptance during qualification testing. Performance trend analysis, thermal shock, random vibration, and accelerated burn-in are the reliability improvement tests that are explained with examples in Chap. 6.

No test sequence or pattern of testing can guarantee detection of all deficiencies [3]. Organizing failure data and identifying the root causes of electronic component failures are presented in Chap. 7. Approaches for analyzing component failures caused by design errors, manufacturing processes, and defective components are explained.

Reliability database system is essential for integrating reliability techniques concurrently with the design and development activities of electronic equipment. The expertise of QA is made available to circuit designers virtually through the system. Adequate information and guidance are presented for designing the reliability database system in Chap. 8.

Cause analysis and implementing corrective actions for failures are fundamental requirements of analyzing field failure data. Additional engineering benefits could be realized by analyzing accumulated field failure data statistically. Analyzing the failure data of field replaceable electronic items using statistical probability distributions and predicting spares are illustrated with examples in Chap. 9.

The method of conducting life test on switches and electromechanical relays to establish the application reliability of the components for nonstandard loads is explained in Appendix A. The concept of Reliability Growth Testing (RGT) is explained in Appendix B. Computing G_{rms} from Acceleration Spectral Density (ASD) levels for random vibration profiles is explained in Appendix C.

References

1. Natarajan, D.: A Practical Design of Lumped, Semi-lumped and Microwave Cavity Filters. Springer, Berlin (2013)
2. Bot, Y.: Improving product's reliability by stress de-rating and design rules check. In: IEEE Proc. Ann. Reliability & Maintainability Symp., (2013): 978-1-4673-4711-2/13
3. Deppe, R.W., Minor, E.O.: Reliability enhancement testing (RET). In: IEEE Proc. Ann. Reliability & Maintainability Symp., pp. 91–98 (1994)

Acknowledgments

Throughout my career, I was assisted by a team of committed and talented engineers and technicians with a lot of enthusiasm and initiatives. I would like to thank the team first for making me a better professional with time.

I would like to thank my wife, Rameswari, for encouraging me to write the book after my retirement. I thank my son, Dr. Kumar Ph.D. for reviewing the manuscript with patience and providing me with valuable suggestions. I also thank my son-in-law, Gopinathan, Semiconductor designer, for technical support.

Most importantly, I thank Krishnamurthy C., Semiconductor professional, General Manager (Retd.), Bharat Electronics, who provided me with valuable opportunities for me to become a true professional in the field of Q&R.

Contents

About the Author

Dhanasekharan Natarajan, Electronics engineer from College of Engineering, Guindy, Chennai in 1970, obtained his post-graduate in Engineering Production (Q&R Option) from the University of Birmingham, UK in 1984. He is a Senior Member (00064352-Member for 27 years), IEEE (MTT Soc.; Rel. Soc.) and his biography is published by Marquis, USA in their fourth edition, "Who's Who in Science & Engineering".
He retired as Assistant Professor in RV College of Engineering, Bangalore. His earlier professional achievements at Bharat Electronics and Radiall Protectron include Application of reliability techniques for defense equipments, Root cause analysis on electronic component failures, Qualification testing of electronic components as per US and Indian military standards, Designing and implementing computerized Quality Management System, Designing software for Optical interferometer, and Design & manufacturing of lumped, semi-lumped and microwave cavity filters using self-developed software. He has authored the book, *A Practical Design of Lumped, Semi-lumped and Microwave Cavity Filters*, Springer, 2013.

Chapter 1
Overview of Reliability Design

Abstract The design phase of electronic equipment has two distinct processes, namely design and development. Circuit design and selection of components are the engineering activities in the design process. The reliability techniques, de-rating, reliable application of components, and FMEA are integrated with the activities of design process. Testing of proto-models is the prime activity in the development process. Proto-models are subjected to overstress analyses, internal electrical, and environmental testing superimposed with reliability improvement tests, and qualification testing. Brief explanation for the reliability techniques and methods of integrating the techniques with the design and development of an electronic item are presented.

1.1 Design and Development of Electronic Equipment

Electronics dominates applications such as entertainment, industrial controls, medical, military, and space. The application requirements of electronic equipment are converted into a set of manufacturing drawings in the design phase of the equipment. The design phase of equipment has two distinct processes, namely design and development. It is explained for a Commercially-Off-The-Shelf (COTS) item, RF Oscillator.

The engineering activities in the design process of RF Oscillator are:

- Deciding functional blocks
- Circuit design for the functional blocks
- Selecting components

The functional blocks of the RF Oscillator is shown in Fig. 1.1. The output of design process is the set of provisional engineering documents.

The activities in development process are manufacturing proto-models of RF Oscillator as per the provisional drawings, evaluating the proto-models, and updating the provisional set of engineering documents after the evaluation of the

© Springer International Publishing Switzerland 2015
D. Natarajan, *Reliable Design of Electronic Equipment*,
DOI 10.1007/978-3-319-09111-2_1

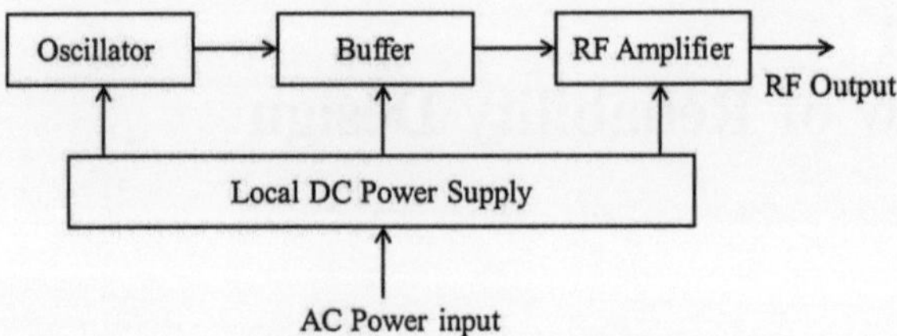

Fig. 1.1 Functional blocks of RF oscillator

proto-models. The outputs of development process are the proven proto-models and the set of finalized engineering documents for the manufacture of RF Oscillator.

1.2 Engineering Reliability in Design and Development

Engineering reliability is defined as integrating reliability techniques concurrently with the activities of design and development of processes of equipment. The integrated reliability design approach minimizes re-design efforts and enhances customer confidence. Reliability techniques for design and development processes differ. The application of reliability techniques are explained for the design and development processes of the RF Oscillator shown in Fig. 1.1 and are applicable for electronic equipment also.

1.2.1 Reliability Techniques for Design Process

Three reliability techniques are relevant for the design process of RF Oscillator as shown in Fig. 1.2. Two techniques are integrated with the selection of components and one technique is integrated with the design of functional blocks. The reliability techniques are:

Selection of components:

(i) De-rating

Adequate de-rating is applied in the selection of components to ensure that the operating electrical and thermal stress levels are less than the maximum ratings of the components.

(ii) Reliable application of components

Even with adequate de-rating, components could fail due to improper applications in circuits. Information provided in the application notes of component manufacturers should be adhered for ensuring the reliable application of components.

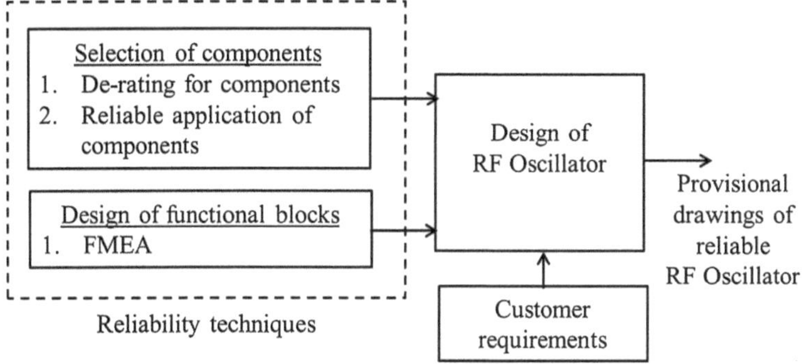

Fig. 1.2 Integrating reliability techniques in design process

Design of functional blocks:

(iii) Failure Modes and Effects Analysis (FMEA)
Failure Mode and Effects Analysis (FMEA) prevents the occurrence of performance failures in the RF Oscillator. Engineering analysis is performed for the assumed failure modes of the modules of RF Oscillator. For example, if the failure mode, poor frequency stability, were to occur for the Oscillator module shown in Fig. 1.1, the failure effect at the output of the RF Oscillator would be 'High phase noise'. Preventive actions are planned to eliminate the failure effect in the circuit design of Oscillator module for ensuring frequency stability.

1.2.2 Reliability Techniques for Development Process

Unlike design process, proto-models exist in the development process of RF Oscillator and hence the reliability techniques are in the form of analyses and tests on the proto-models. The analyses and tests verify the effectiveness of applying the reliability techniques in the design process of RF Oscillator. The reliability techniques for development process as shown in Fig. 1.3 are:

(i) Overstress analyses:
Electrical and thermal overstress analyses are conducted on the modules and on the RF Oscillator to confirm that components with adequate de-rating are used. Vibration resonance overstress analysis is conducted to identify fatigue related defective components.

(ii) Internal verification testing:
Design errors in the selection and application of components might still exist in the design of equipment. There could be unknown failure modes that were not identified in FMEA. Performance trend analysis, thermal

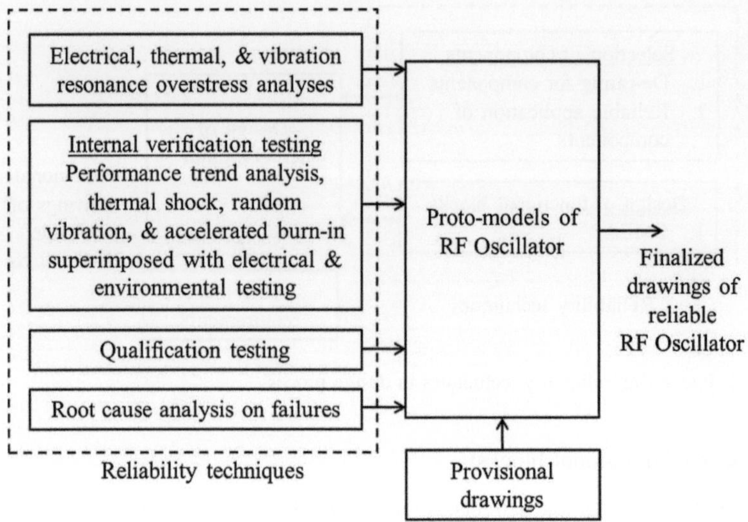

Fig. 1.3 Integrating reliability techniques in development process

shock, random vibration and accelerated burn-in tests are part of internal verification testing to precipitate design errors and to achieve first time acceptance in qualification testing. Some of the tests are superimposed with electrical performance verification and environmental tests to save test time.

(iii) Qualification testing:
Electrical and environmental tests are conducted on the RF Oscillator to verify compliance to contractual specifications.

(iv) Failure analysis:
Component failures might be observed in the development process of RF Oscillator. Root causes of the failures are identified and corrective actions are implemented to prevent their recurrences.

1.3 Computerized System for Reliability Design

Engineering information is required for applying reliability techniques concurrently with the activities of design and development of equipment. The information is distributed in the catalogs and application notes of component manufacturers, military standards, and literature. If right information is not available at the right time, the concurrent design efforts might be compromised. The best way to channelize engineering information is to establish computerized reliability data base system. The reliability database system for design and development is shown in Fig. 1.4.

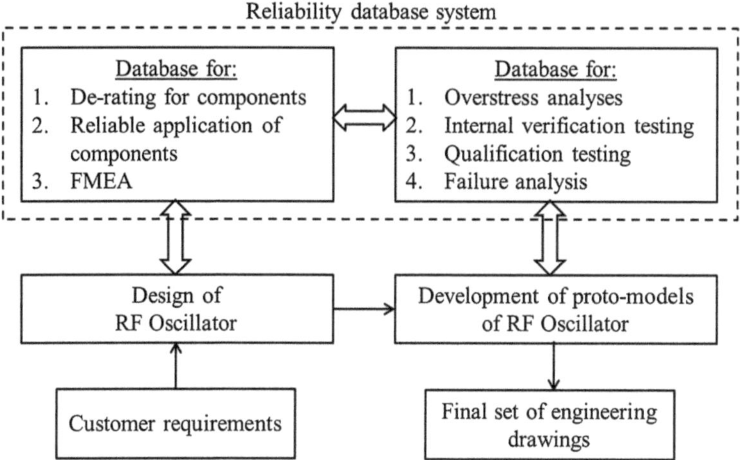

Fig. 1.4 Reliability database system for design and development

The database system provides the information instantly for circuit designers for implementing concurrent design approach in the design and development of equipment. The reliability techniques, the application of the techniques, and the design of reliability database system are explained in the subsequent chapters, focusing the needs of electronic equipment and COTS items.

Chapter 2
Generic Stress Factors and De-rating

Abstract The data sheets of components specify maximum ratings and electrical characteristics. The parameters of maximum ratings are the reliability-related stress factors of electronic components. The generic stress factors, ambient temperature, surface temperature, and junction temperature that are common to most components are explained with examples. The basics of de-rating, applying de-rating for stress factors, and thermal resistance are also explained.

2.1 Introduction

Components are the basic elements for developing electronic equipment, and the data sheets of manufacturers are used for selecting the components. The data sheets specify maximum ratings and electrical characteristics for components. The parameters of maximum ratings are the reliability-related stress factors. Examples of stress factors are rated voltage, power dissipation, maximum junction temperature, etc. Exceeding the maximum values of the stress factors reduces the reliability of components. The reliability technique, de-rating, is applied for stress factors in the selection of components to ensure that the operating stress levels of the components are within their maximum ratings.

The list of stress factors for the categories of electronic components is presented in Chap. 3. The generic stress factors that are applicable to most components and the basics of de-rating are explained with examples. Thermal resistance is not a stress factor but it is also explained as it is used for computing the junction temperature of semiconductor devices.

2.2 Ambient Temperature

Ambient temperature (T_A) of a component is the temperature of air in the vicinity of the component. The internal ambient temperature of components is usually higher than the external ambient temperature of equipment due to heat dissipation

© Springer International Publishing Switzerland 2015

D. Natarajan, *Reliable Design of Electronic Equipment*,

DOI 10.1007/978-3-319-09111-2_2

in the equipment. It is one of the critical inputs for selecting electronic components. Three examples are provided for measuring the internal ambient temperature of components.

Example-1:

Assume that the power dissipation of an assembled PCB is low and the
temperature rise is negligible, i.e., less than 5 °C. The surface temperature
of PCB is the ambient temperature for the components of the PCB.

Example-2:

The surface temperature of the metal enclosure of plug-in module is the
ambient temperature for the components mounted on the PCB of the
module.

Example-3:

Assume a transistor is mounted in the vicinity of a heat dissipating component such as wire-wound resistor. The surface temperature of the wire-
wound resistor is approximately the ambient temperature for the transistor.

2.3 Surface Temperature

Surface temperature is specified for resistors and it is the temperature at the
external area of the resistors. Case temperature instead of surface temperature is
used for transistors, packaged in metal cases such as TO3 (Transistor Outline-3).
Case temperature is also relevant for ceramic and plastic encapsulated ICs. The
surfaces of components are probed by thermocouples to measure the highest
temperature. The surface or case temperature of components is related to the
power dissipation in the components.

2.4 Thermal Resistance

The electrical resistance of a resistor is given by the ratio of voltage gradient
between the terminals of the resistor to the current flowing through the resistor.
Thermal resistance is analogous to electrical resistance. Temperature gradient
exists between two points in thermal conduction path. Thermal resistance is the
ratio of temperature gradient to the power dissipation of heat source. An example
is provided for better understanding.

A power transistor with a heat sink is shown in Fig. 2.1. Heat is generated by
power dissipation in the transistor. Conduction is the primary means of heat

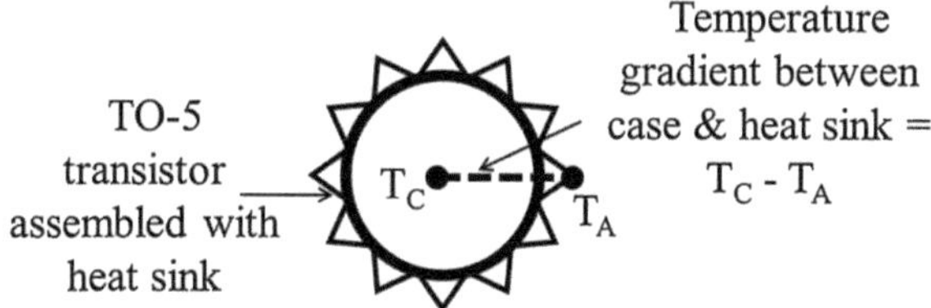

Fig. 2.1 Temperature gradient in power transistor

dissipation from the transistor to ambient, and the path of conduction is through the heat sink. Temperature gradient exists in the thermal conduction path between the case of the transistor and the heat sink and it is shown in Fig. 2.1. Assume the following:

P_D: Power dissipated by the transistor in watts
T_C: Case temperature of the transistor
T_A: Temperature at the fins of the heat sink, i.e., ambient temperature

Thermal resistance of the heat sink, $(\theta_{c-a}) = (T_C - T_A)/P_D$

The unit of thermal resistance is °C/W. Using heat sinks having lower thermal resistance reduces the case temperature of the power transistor. Increasing dimensions or the number of fins or using better material reduces the thermal resistance of the heat sink.

2.5 Junction Temperature

Power dissipation in semiconductor devices increases the operating junction temperatures of the devices. The heat generated at the junction of semiconductor devices is dissipated to ambient thorough the case and leads of the devices. The construction of the devices is such that the heat at the junctions is conducted through conducting epoxy, lead frame, etc., before reaching the case or package of the devices. The conduction path results in temperature gradient between the junction and case of the devices. The temperature gradient is proportional to power dissipation and the ratio is specified as thermal resistance (°C/W) between junction and the case (θ_{j-c}) of semiconductor devices by manufacturers. Operating junction temperature is one of the critical parameters of semiconductor devices as it decides the reliability of the devices. Maximum junction temperature for semiconductor devices should never be exceeded in the application of the devices in circuits. θ_{j-c} is used for computing the operating junction temperature of the devices in circuits.

2.5.1 Computing Junction Temperature

The steps in computing the operating junction temperature of a semiconductor device are:

(i)   Determine the power dissipation (P_D)
(ii) Estimate or measure the case temperature (T_C)
(iii) Obtain θ_{j-c} from the data sheet
(iv) Compute operating junction temperature:
$$T_j = T_C + \theta_{j-c} * P_D$$

Heat sink might be used for semiconductor devices to reduce operating junction temperature. The thermal resistance of the heat sink and the temperature of the heat sink (T_A) are used for computing the junction temperature.

Junction temperature, $T_j = T_A + (\theta_{j-c} + \theta_{c-a})P_D$

Data sheets might specify thermal resistance from junction to ambient, θ_{j-a}, for some devices. Operating junction temperature of the devices is computed from ambient temperature.

Junction temperature, $T_j = T_A + \theta_{j-a} * P_D$

2.6 Basics of De-rating

De-rating is the ratio of operating stress level to the rating (strength) of component. Components with adequate strength should be selected to withstand the estimated operating stress levels. Consider the selection of a resistor. Power dissipation is one of the stress factors for a resistor. Let the estimated power dissipation, i.e., the operating stress level of the resistor in a circuit be 0.47 W. A resistor with 1 W rating (adequate strength) should be selected although a resistor with 0.5 W rating (inadequate strength) is available. De-rating is usually expressed in percentage. The de-rating level for the resistor is [100*(1 – 0.47 W)/1 W], i.e., 53 %. It is quite possible that the stress levels of some components could not be assessed while selecting components. In such cases, operating stress levels on the components are measured on the proto-models of equipment as explained in Chap. 5 and the selection of the components are reviewed. De-rating is applicable for both electrical and thermal ratings of components.

2.6.1 Need for De-rating

Electronic components fail during design, manufacturing, and field usage of equipment for various reasons. The nature of the failures of components could be classified as either mechanical or chemical [1]. The coil ends of electromagnetic relays are soldered internally to the terminals of the relays. The snapping of the coil ends from the terminals under mechanical shock impacts is an example of mechanical failure. Reduction in the gain of a power amplifier chip is an example of chemical failure, caused by the chemical reaction on the die of the chip.

As per Arrhenius model, the rate of chemical reaction increases with temperature and it doubles approximately with every 10 °C increase in temperature [2]. Ambient temperature and electrical stresses increase the operating temperature of components. Applying de-rating for the stress factors of components reduces the operating temperature of components thereby reducing the failure rate of components considerably.

2.6.2 Examples of De-rating

De-rating for the stress factors except power dissipation of components could be applied directly as shown in the examples below. De-rating for power dissipation is explained in Sect. 2.6.2.1.

Example-1:

Absolute maximum junction temperature for transistor: 150 °C max.
De-rating level: 20 %
Circuit design limits operating junction temperature to 120 °C.

Example-2:

Rated voltage of chip ceramic capacitor: 50 V DC max
De-rating level: 10 %
Circuit design limits operating voltage to 45 V DC.

2.6.2.1 De-rating for Power Dissipation

Maximum rated power dissipation is normally specified at 25 °C case or ambient temperature for semi-conductor devices and at 70 °C ambient temperature for resistors. Beyond the standardized temperature of rating, the rated power of the components decreases. Power rating versus temperature graphs for a semiconductor device and a resistor are shown in Fig. 2.2a, b respectively. Manufacturers refer the graphs as de-rating or re-rating graphs. As the graphs indicate the power rating of components with temperature, it is appropriate to refer the graphs as re-rating graphs.

Power rating at the operating temperature of components is computed from the re-rating graphs and de-rating is applied for the computed power rating. The computations of power rating and de-rating the rated power are explained for a transistor. The procedure is applicable for resistors also.

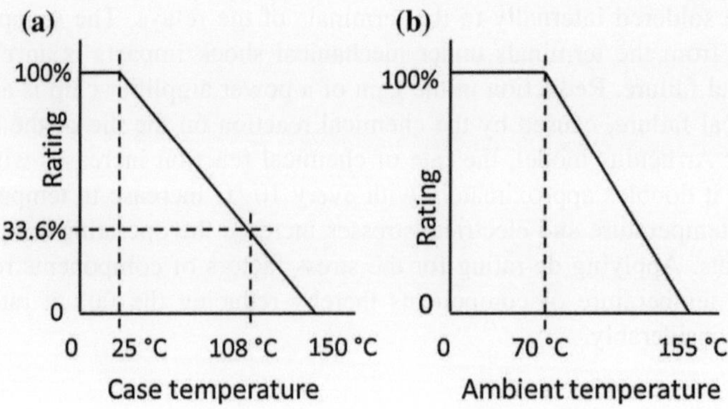

Fig. 2.2 Power rating with temperature—examples, **a** transistor rating **b** resistor rating

Example

The rated power of a transistor is 30 W at 25 °C case temperature. The power re-rating with temperature, shown in Fig. 2.2a is specified by the manufacturer of the transistor. Compute the power handling capacity of the transistor with 20 % de-rating level at 108 °C case temperature.

Solution

Rated power of transistor at 25 °C case temperature: 30 W
Operating case temperature of the transistor: 108 °C
Rated power of the transistor at 108 °C from Fig. 2.2a = 33 % of 30 W = 10 W
De-rating level: 20 %
Power handling capacity at 108 °C = 10 W − (20 % of 10 W) = 8 W

A linear equation in the form, $y = m * x + c$, could be fitted for the re-rating graphs for determining rated power with temperature, where m is the slope of the straight line characteristic.

Assessment Exercises

1. Say True or False:

   (i) De-rating means that applied stress should not exceed strength.
   (ii) The terms, de-rating and re-rating, mean the same.
   (iii) Temperature gradient between junction and case of transistor increases with power dissipation in the transistor.
   (iv) Junction temperature and case temperature of a transistor are equal if thermal resistance between junction and case is zero.

   (v) Thermal resistance between junction and case of transistor increases with power dissipation in the transistor.

2. A wire-wound resistor, embedded in metal case, is dissipating 18 W in a module. The operating ambient temperature of the resistor is 55 °C. The thermal resistance between the surface of the resistor and the metal case is 1.2 °C/W and that between the metal case and ambient is 5.8 °C/W. Compute the surface temperature of the resistor.

   If the maximum surface temperature of the resistor is 200 °C, does the application satisfy the de-rating level of 10 %?

3. A low power transistor in TO-5 case is mounted in a PCB. The absolute maximum ratings and thermal resistance of the transistor are:

   Power dissipation at $T_A = 25$ °C: 0.8 W

   Junction temperature, T_{j-max}: 175 °C.

   Thermal resistance, junction-ambient (θ_{j-a}): 190 °C/W.

   The power rating curve with ambient temperature is shown below.

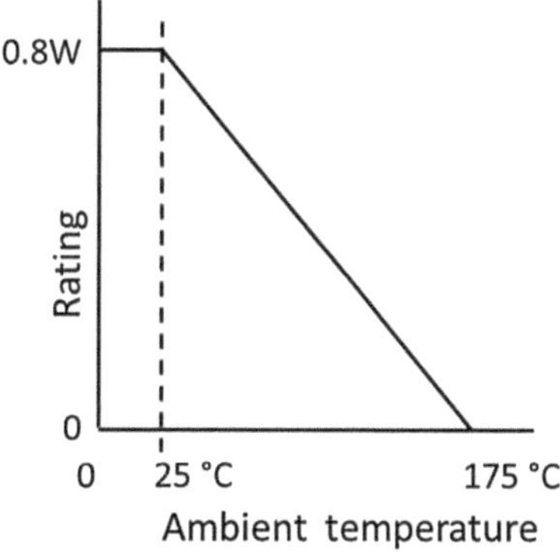

   Compute the maximum power that could be dissipated by the transistor with 15 % de-rating assuming the internal operating ambient temperature of the PCB is 87 °C.

References

1. Sherwin, D.: Lectures notes, Engineering Production Department, University of Birmingham, UK (1983–1984)
2. Mortimer, M., Taylor, P.G. (eds.) The Chemical Kinetics and Mechanisms. Royal Society of Chemistry publishing, Cambridge, UK (2002)

Chapter 3
Selection and Application of Components

Abstract High level of standardization in electronic components simplifies the process of selecting reliable components. The reliability needs of components could be summarized as applying de-rating to stress factors and ensuring proper application. The stress factors of ICs, semiconductor devices, and passive components are listed. Guidelines for deciding the quantum of de-rating for the stress factors are provided. Application information for the components is also presented and they are based on the application notes of manufacturers, literature, and root cause analysis on component failures.

3.1 Basis of Selection

Electronic components are generally selected after finalizing circuit design. The broad categories of electronic components are integrated circuits (ICs), discrete semiconductor devices, and passive components. The list of the components is quite large. The critical factors for selecting components are shown in Fig. 3.1. Performance and cost requirements are specific to electronic equipment. Reliability needs of components are explained as they are common to equipment. Addressing the reliability needs in the selection of components prevents component failures.

3.2 Reliability Needs of Components

High level of standardization in electronic components provides unique reliability advantage. The materials, finish, dimensions (outline), and characteristics of most electronic components are well established and controlled by military and other international standards. The methods of conducting electrical and environmental testing on the components are also controlled by the standards. Hence, it would be

© Springer International Publishing Switzerland 2015
D. Natarajan, *Reliable Design of Electronic Equipment*,
DOI 10.1007/978-3-319-09111-2_3

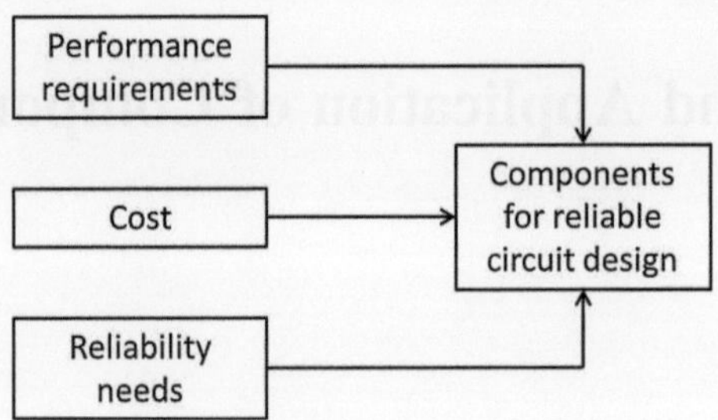

Fig. 3.1 Critical factors for selecting components

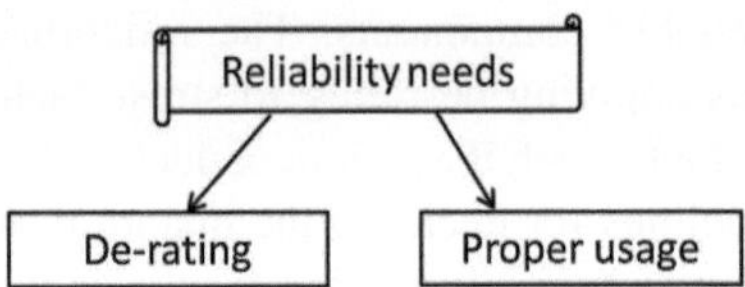

Fig. 3.2 Reliability needs of components

adequate to address the reliability needs shown in Fig. 3.2 when selecting components for the design of electronic Equipment and COTS items. The needs are also emphasized in the application notes of component manufacturers. The reliability needs of components are:

(i) Applying de-rating to the stress factors of components
(ii) Ensuring proper usage, i.e., the application of components

3.2.1 De-rating for Stress Factors

In civil engineering designs, applying stress less than strength is termed as *factor of safety*. In mechanical engineering designs, the term, *safety margin* [1] is used. Factor of safety and safety margin are designed to be quite high, i.e., the applied stress could even be less than 10 % of strength of materials considering the need for the high reliability of structures for a long period of time and uncertainties in design. In the design of electronic Equipment, excessive de-rating need not be applied as the intrinsic reliability of electronic components is high due to technological advancements. The quantum of de-rating for stress factors is decided considering the needs of components, circuits, and the application of Equipment. Guidelines for deciding de-rating levels are:

(i) Irrespective of the application needs of components and Equipment, a minimum level of 10 % is recommended for de-rating;

(ii)   Higher level of de-rating might increase the cost and size of components but it should be accepted and applied for:

- Components used in circuits having high transient (surge) voltages and currents (Ex. Power supply modules). For example, de-rating level of 17 % is applied for the IC, DC-DC Converter [2].
- Components such as RF power ICs, RF power transistors, and power MOSFETs due to uncertainties in their applications. For example, de-rating level of 27 % is recommended for the application of RF Power transistor [3].
- Components used in mission critical Equipment (Ex.: space and military)

(iii) De-rating should not be applied for stress factors if the performance reliability of components is affected. For example, de-rating the coil voltage of electromagnetic relays is not recommended. The relays are current operated devices. They might not operate with de-rated coil voltage at high temperature due to the reduction of coil current caused by the increase in coil resistance.

Considering de-rating alone is not adequate. Proper usage or the application of components should also be considered in the selection of components.

3.2.2 Application of Components

Design errors in the application of components are one of the causes of component failures. Most products seldom fail due to part failure, but often due to incorrect application and integration of those parts [4]. Two examples illustrate improper application of components and the preventive actions in the selection of the components.

Assume the end product is high power SMPS. The maximum operating temperature of the SMPS is specified as 65 °C. Selecting components for the ambient temperature of 65 °C with proper de-rating is an example of improper application. The induced temperature rise within the SMPS is not considered. The design of the SMPS is marginal and the components might fail. The induced temperature rise should be estimated and added to 65 °C for computing the internal ambient temperature of the SMPS. Components for the SMPS should be selected with proper de-rating for the internal ambient temperature.

The second example is regarding the application of electromagnetic relay, although the relay is not popular in digital designs. Consider an electromagnetic relay with 1 A contact rating. If the relay needs to handle a load current of 1.6 A, paralleling two contacts of the relay is an improper application. The contacts of the relay do not close simultaneously due to contact bounce and the contact that closes first switches 1.6 A reducing relay reliability. A relay with 2 A contact rating should be selected for handling 1.6 A and the relay satisfies the requirements of both de-rating and satisfactory application.

3.3 Stress Factors and Application Information

The list of stress factors for de-rating and a sample of application information are presented for ICs, discrete semiconductor devices, and some of the widely used passive components. The stress factors applicable for a specific component should be identified by designers from the list using the data sheet of the component. The application information is not exhaustive but it sets the lead for collecting additional information. The sources of the information are:

  (i)   Technical data sheets of components
  (ii) Application notes of component manufacturers
 (iii) Military Standards for components
  (iv) Literature survey
  (v) Learning from root cause analysis on component failures

3.3.1 Integrated Circuits

The stress factors and application information are presented for general purpose ICs. As the level of integration is ever-increasing and the rate of obsolescence is high, the collection of application information should focus on the ICs needed by organizations.

3.3.1.1 Stress Factors for De-rating

- Supply voltage
- Supply current
- Input voltage
- Input current
- Output current
- Power dissipation
- Maximum junction temperature, $T_{j(max)}$
- Operating temperature range

3.3.1.2 Application Reliability Information

- It is preferred to operate ICs with the supply voltage recommended by manufacturers to ensure electrical performance reliability.
- Thermal resistance varies with the type of package of integrated circuit. The maximum value of thermal resistance should be used for computing junction temperature. Nominal values should not be used for the computation.
- Improved thermal performance could be realized by direct mounting of integrated circuits in PCBs instead of socket mounting [5].
- For integrated circuits with negligible power input and power output such as digital gates, power dissipation (P_D) is the product of supply voltage (V_{CC}) and supply current (I_{CC}). For power ICs, input power and output power should also be included for calculated power dissipation [6].

$$P_D = V_{CC} * I_{CC} + \text{Power input} - \text{Power output}$$

For RF power ICs, there could be reflected power from mismatched load and this would add to power dissipation as the reflected power is dissipated in the IC.

- The existence of latch-up induced electrical overstress should be examined and prevented. Latch-up is a potentially destructive situation in which a parasitic active device is triggered, shorting the positive and negative supplies together. Latch-up is triggered if the input or output pin voltage is raised above the positive supply voltage or lowered below the negative supply voltage [7].

3.3.2 Discrete Semiconductor Devices

The stress factors and application information are presented for general purpose diodes and transistors. Like ICs, the collection of application information should focus on the semiconductor devices needed by organizations.

3.3.2.1 Stress Factors for De-rating

- Current rating
- Reverse voltage
- Junction breakdown voltages
- Power dissipation
- dv/dt rating for power MOSFETs
- Maximum junction temperature, $T_{j(max)}$
- Operating temperature range

3.3.2.2 Application Reliability Information

- When a power transistor is subjected to pulsed or intermittent load, higher peak power dissipation is permitted [8].
- Although the MOSFETs are operating within the absolute maximum ratings of voltages, currents, power dissipation, and junction temperature, the devices could still fail due to transient drain voltage rise with time (dv/dt) under specific load conditions such as unclamped inductive switching and motor loads or in circuits that use parasitic body diode [9].
- Operating junction temperature of SOT-223 power MOSFET could be reduced by 10–15 % by optimizing the design of copper mounting pads [10].
- Microwave transistors used in amplifiers require specially designed feedback networks to ensure bias stability as the conventional emitter bypass capacitor gives rise to bias oscillations degrading noise figure [11].

3.3.3 Resistors

Carbon composition, metal oxide, metal film, and cermet resistors and wire wound resistors are commonly used in electronic Equipment. Fixed resistors are available in both chip and leaded form. Variable resistors are available in single and multiturn configurations. The stress factors are common to all the types of resistors.

3.3.3.1 Stress Factors for De-rating

- Power dissipation
- Limiting voltage
- Maximum surface temperature
- Operating temperature range

3.3.3.2 Application Reliability Information

- Embedded resistors have metal cases over the core of the resistors. In such resistors, the operating core temperature is maintained below the rated temperature. The core temperature is determined using the measured case temperature and thermal resistance between core and case [12].
- Limiting voltage (rated voltage) rating is related to the physical size of resistors, i.e., the creep distance for arcing between the terminals. For example, limiting voltage rating of 1 W resistor is higher than that of 0.5 W resistor. Limiting voltage is a stress factor in the selection of resistors with resistance values higher than critical value of resistance for voltage divider circuits and in high voltage power supplies. Critical value of resistance is defined as the value of resistor which dissipates rated power at its limiting voltage.

- Variable resistors are used to adjust voltage or current in circuits. The minimum value of resistance is realized when the wiper contact is at the extreme end of the track and it is termed as end resistance. The circuit design should ensure that the current through end resistance does not exceed the rated current of variable resistors.
- Variable resistors with palladium silver plated internal contacts are more reliable than silver plated contacts.

3.3.4 Capacitors

Capacitors with ceramic dielectrics and electrolytic capacitors are commonly used in electronic Equipment. There are many styles in each type of capacitor. For example, ceramic capacitors are available with single layer and multilayers in chip form and leaded form. Aluminum electrolytic, solid aluminum, and tantalum are examples of electrolytic capacitors.

3.3.4.1 Stress Factors for De-rating

- Voltage
- Ripple current (for electrolytic capacitors)
- Operating temperature range

3.3.4.2 Application Reliability Information

- Electrolytic capacitors might be required to be connected in series in applications such as AC drive systems for industrial applications. Variations in leakage current could lead to exceeding the rated voltage of capacitors, compromising the reliability of the capacitors [12].
- Both transient and steady-state ripple current through solid tantalum capacitors should not to exceed the rated ripple current of the capacitors [13].
- Solid aluminum capacitors, i.e., polymer capacitors have the advantages of aluminum electrolytic capacitors and have very low ESR (Equivalent Series Resistance) and ESL (Equivalent Series Inductance), making them suitable for operations at high temperature and frequency [14].
- Using bonding pads larger than the size of chip ceramic capacitor terminations ensures higher reliability for the capacitors [15].
- Dual voltage rating is specified for a few types of capacitors such as solid tantalum and polyester capacitors. For example, voltage rating is specified as 25 V DC max up to 85 °C and it is 20 V DC max up to 125 °C. De-rating should be applied to rated voltage applicable at the operating temperature.
- In high-power RF applications, the equivalent series resistance (ESR) becomes significant causing rise in the case temperature of chip capacitors [16, 17]. Larger case size capacitors could handle higher RF power.

3.3.5 Inductors and Transformers

Inductors and transformers are generally reliable provided hot spot temperature of winding wire is not exceeded. Maximum hot spot temperature of winding wire is specified by manufacturers. The temperature depends on the type and the thickness of covering over copper.

3.3.5.1 Stress Factors for De-rating

- Hot spot temperature for power inductors and transformers
- Voltage
- Current
- Power dissipation
- Operating temperature range

3.3.5.2 Application Reliability Information

Power dissipation increases the temperature of power transformers and inductors resulting in hot spots on winding wires. Standard graphs in military standards and application notes for transformers could be used for determining hot spot temperature.

3.3.6 Connectors

Large types of connectors are available for the design of electronic Equipment. Broadly, connectors could be categorized as RF coaxial connectors and general purpose connectors. Examples of RF coaxial connectors are BNC, N, TNC, and SMA series. Examples of general purpose connectors are module connectors and circular connectors. Module connectors are available in pin and shell polarized configurations. Circular connectors with threaded and bayonet couplings are available.

3.3.6.1 Stress Factors for De-rating

- Rated voltage
- RF power rating with frequency (for coaxial connectors)
- Rated current
- Operating temperature range

3.3.6.2 Application Reliability Information

- Coaxial cables are assembled to the cabling type plugs and sockets of RF coaxial connectors. Cabling a large size connector with a very low diameter coaxial cable could cause the failure of the cable assembly due to poor cable retention. For example, cabling RG316/u cable (Φ3 mm) to N plug is not recommended especially for repeated engagement and disengagement.
- Cabling type coaxial plugs and sockets are available with captivated or non-captivated center contacts. The connectors with captivated center contacts ensure reliable operation at temperature extremes and for repeated engagement and disengagement.
- Silver plated coaxial connectors generate low intermodulation distortion compared to nickel plating. Connector manufacturers have developed alternative finishes to overcome the possibilities of tarnishing in silver plated connectors. Radiall's BBR (Bright Bronze Radiall) and GBR (Gold Bronze Radiall) plating and other equivalent plating technologies have the advantages of silver plating but without tarnishing effects.
- Manufacturer recommended procedures and tools should be used for the cable assembly of coaxial connectors to ensure high RF performance and cable retention.
- Mechanical adapter should be used for mating 50 Ω N-plug with 75 Ω N-Socket to prevent damaging the female contact of the 75 Ω N-Socket.
- Coaxial connectors are available for clamping or crimping coaxial cables to the connectors. Crimped cable assemblies have higher cable retention compared to clamped cable assemblies and they are more reliable for repeated engagement and disengagement.
- Electrical overstress occurs if two contacts in general purpose connectors are paralleled for doubling the rated current since the current divides proportionately as per contact resistance. Paralleling of general purpose connector contacts for the purpose of carrying more than the rated current of contacts should be done after statistically analyzing the worst-case variation in the contact resistance of connector contacts.

3.3.7 Switches

Very large types of mechanical switches are available for the design of electronic Equipment. Examples of switches are toggle switches, push-button switches, rotary switches, micro switches, slide switches, rocker switches, and membrane switches. There are many styles in each type of switches. For example, SPDT (Single Pole Double Throw), DPDT (Double Pole Double Throw), and panel sealed are some of the styles in toggle switches. Rotary switches are available with many circuits and ways and make before break or break before make configurations. Slide switches, rotary wafer switches, and membrane switches are used signal switching applications and the other types of switches are used for both signal and power switching applications. Solid-state coaxial switches are available for RF signal switching applications.

3.3.7.1 Stress Factors for De-rating

- Rated voltage
- Rated current
- Operating temperature range

3.3.7.2 Application Reliability Information

- Military specifications and most manufacturers specify contact rating at the standardized voltages, 28 V DC and 120 V AC for switches. Unless otherwise specified, the ratings are applicable for resistive loads. Current ratings for inductive and motors loads are lower than resistive load ratings due to high transients associated with the loads. If the load ratings of contacts are not specified for inductive and motor loads in data sheets, application notes of manufacturers and military specifications should be referred for deciding the ratings.
- Circuit application needs might exist to switch nonstandard loads. For example, an application might need to switch a resistive load of 48 V/1A DC using a toggle switch. Very few manufacturers provide supplementary data for switching nonstandard loads. Supplementary information provided by NKK, Japan could be used to select switches initially. A toggle switch with 28 V/5A rating could be selected for 48 V/1A resistive load switching. The application reliability of the switches for nonstandard loads should be confirmed by conducting life test on the toggle switch as described in Appendix A.
- Some circuits require switching very low currents of the order of a few tens of microamperes and they are termed as low-level circuits. Switches with suitable finishes such as gold plating for contacts are reliable for low-level circuits as they prevent the formation of oxide layer during usage.

3.3.8 Relays

Unsealed general purpose relays, hermetically sealed relays, thermal relays, reed relays are some of the examples of electromechanical relays. Solid date relays are popular in digital circuits. SPDT (Single Pole Double Throw) and DPDT (Double Pole Double Throw) are some of the styles in relays.

3.3.8.1 Stress Factors for De-rating

Electromechanical relays:

- Rated coil voltage
- Rated contact voltage
- Rated contact current
- Operating temperature range

Solid-state relays:

- Junction temperature
- Input current
- Reversed input voltage
- Input power
- Output power
- Output current
- Output voltage
- Operating temperature range

3.3.8.2 Application Reliability Information

Electromechanical relays:

- Like switches, military specifications and most manufacturers specify contact rating at the standardized voltages, 28 V DC and 120 V AC. The application reliability information of switches could be referred for deciding the ratings for inductive, motor, and nonstandard loads.
- Relays are current operated devices and hence de-rating should not be applied for rated coil voltage. Operating a relay with coil voltage less than rated voltage causes unreliable operation at high temperature.
- Relay contacts should not be paralleled for the purpose of doubling the rating of the contacts. The contacts of the relay do not close simultaneously due to contact bounces and the relay fails due to electrical overstress on the contacts.
- Minimum spacing between relays should be planned for multiple mounting of relays to avoid interaction between the magnetic fields of the relays [18].
- Surge suppressor diode is used across relay coil to suppress the reverse high voltage (back emf) generated during the switching of the relay. The suppression diode should be as close to the coil as possible [18].

Solid-state relays:

- The contacts of solid-state relays might be exposed to destructively high voltages in switching circuits. The contacts should be protected by using metal oxide varistors or transient absorbers [19].

- For solid-state relay that uses MOSFETs at its output, operating junction temperature should be computed to ensure that $T_{j(max)}$ of the MOSFET is not exceeded [20]. De-rating for the operating junction temperature of the MOSFET should also be applied.
- Transient current handling capability of solid-state relays are much higher than electromechanical relays [20].
- If solid-state relay is turned on inadvertently by the leakage current of driver circuit, it could be prevented by placing a shunt resistor in parallel with LED [21].

3.3.9 Fuses

Fuses protect circuits that are connected after the fuses under overload conditions. Quick acting and time lag fuses are available. The time-current characteristics of the fuses differ and the data sheets of manufacturers could be referred for details. Quick acting fuses are suitable for resistive load circuits. Time lag fuses are suitable for capacitive circuits, which are characterized by power-on inrush current.

3.3.9.1 Stress Factors for De-rating

- Nominal current
- Rated voltage
- Breaking capacity
- Nominal melting (A^2 s)
- Operating temperature range

3.3.9.2 Application Reliability Information

- Most manufacturers of fuses recommend higher de-rating, i.e., operating the fuses at lower than the specified nominal current (I_n). In general, operating at no more than 75 % of nominal current is recommended [22].
- The nominal current carrying capability of fuses decreases with increase in ambient temperature and the data sheets of fuse manufacturers could be referred for details. Improper selection of fuse holders also contributes to increase in ambient temperature of fuses. Fuse holder with 5 A rating is required for 3 A fuses [22].
- Breaking capacity of fuse is the capability of the fuse to open under short circuit condition in a circuit without endangering its surroundings. Open-circuiting of fuses with cracking or explosion of glass tube or loosening of end caps under short circuits is not acceptable. The maximum availability of short

circuit current in circuits should be estimated. Fuses with breaking capacity exceeding the estimated short circuit current should be selected.

- Fuses are capable carrying higher than specified nominal current (I_n) for a short duration. Pulse current capability is specified as nominal melting in A^2-s by fuse manufacturers. The pulse current energy should be less than the nominal melting energy specified for the fuse and the time lapse between the pulses should be more than 10 s [22]. If the time lapse between pulses is less, it results in the accumulation of heat at the fuse wire element causing premature opening of fuses.
- The ends of fuse wire elements are soldered to end caps. Fuses with axial leads attached to the end caps are available for PCB mounting. Adequate care should be exercised while soldering fuses on to PCBs to prevent reflow of solder from end caps over fuse wire elements. Reflow of solder increases the cross-sectional dimensions of fuse wire element, altering the time-current characteristics of fuses.

3.3.10 Coaxial Cables and PCBs

- Lower operating ambient temperature is recommended for polyethylene dielectric coaxial cables to prevent the degradation of characteristic impedance and attenuation characteristic of the cables.
- The bending diameter of coaxial cables should not be less than ten times the diameter of the cables to preserve their nominal characteristic impedance.
- PCBs are reliable by design for most applications except for additional considerations for those used in power modules. The current rating of PCB conductors (tracks) is decided by the width and thickness of the conductors. The temperature rise of the conductors should be limited to ensure that the operating temperatures of PCBs are well within the maximum temperature of copper clad sheet to prevent degradation of the peel strength of the conductors and the pull-off strength of bond pads.
- The thickness, size, and mounting of PCBs should be designed to prevent vibration resonance of components mounted on the PCBs.

Assessment Exercises

1. Say True or False:

    (i)   De-rating is one of the application reliability requirements of components.
    (ii) Inductive load rating is always lower than the resistive load rating of a toggle switch.

(iii) De-rating is applicable for the stress factor, power rating, of 2.2 MΩ, 1 W, Metal Film resistor.

(iv) Junction temperature of a transistor should be calculated using nominal thermal resistance, mentioned in the catalog of the transistor.

(v) Breaking capacity of a cartridge fuse depends only on the size (dimensions) of the fuse and not on the material of the cartridge.

2. Explain the reliability needs of electronic components with examples.

3. The specified maximum operating ambient temperature of 800 W DC-DC Power Inverter is 70 °C. A designer selects plastic encapsulated ICs having operating temperature range, (-20 to 70 °C) as they are economical and adequate for using them in the Power Converter. Comment on the selection of the ICs for the design of the Inverter.

4. List the guidelines for deciding the quantum of de-rating for the stress factors of components.

5. Explain the considerations for the selection and application of cartridge fuses.

Research Assignments

1. List stress factors and information for reliable application for the ICs listed below with documentary evidences:

   (i) Microprocessor
   (ii) EEPROM
   (iii) LED Display driver
   (iv) Audio power amplifier

2. List the stress factors and information for reliable application for the semiconductor devices and passive components listed below with documentary evidences:

   (i) Rectifier diodes
   (ii) RF Power Transistor
   (iii) MOSFETs
   (iv) Solid-state relay
   (v) Polymer solid aluminum electrolytic capacitors
   (vi) Multilayer ceramic chip capacitors for MW applications
   (vii) Time lag fuse

References

1. Carter, A.D.S.: Mechanical Reliability. John Wiley & Sons, (1986)
2. Application Note of NCP6361, Buck Converter With Bypass Mode for RF Power Amplifiers, NCP6361/D, Oct. 2013-Rev.2, On Semiconductor, USA
3. Data Sheet of TRF8010, RF Transmit Driver, SLWS031B, May 1997, Texas Instruments, USA
4. Albertyn Barnard, Ten Things You Should Know About HALT & HASS, IEEE Proc. Ann. Reliability & Maintainability Symp., 2012, 978-1-4577-1851-9/12
5. Understanding Integrated Circuit Power Capabilities, National Semiconductor Application Note, MS009312, April 2000
6. Thermal Design Considerations for RF Power Amplifier Devices, TI Application Note, SLWA 009, Feb 1998
7. Niall Lyne, Electrically Induced Damage to Standard Linear Integrated Circuits: The Most Common Causes and the Associated Fixes to Prevent Recurrence, Application Note-AN397, Analog Devices
8. Power Semiconductor Applications, Philips Semiconductors, 1994
9. Application Note-AN9010, MOSFET Basics, Fairchild Semiconductor Corporation, Rev 1.0.5, April 2013
10. Application Note-AN1028, Maximum Power Enhancement techniques for SOT-223 Power MOSFETs, Fairchild Semiconductor Corporation, Rev B, Aug. 1998
11. Microwave Transistor Bias Considerations, Microwave Transistor Application Note 944-1, Avago Technologies, USA, 5988-0424EN – May 11, 2010
12. Kaveh, M., Yellamati, D., Goktas, Y.: Reliability Testing, pp. 1–7. Reliability and Maintainability Symposium, Analysis and Prediction of Balancing Resistors (2012)
13. Reed.E.K, Tantalum Chip Capacitor Reliability in High Surge and Ripple Current Applications, Proceedings of 44th Electronic Components and Technology Conference, 1994, pp 861-868
14. Application Note on Conductive Polymer Aluminum Solid Capacitors, Nippon Chemi-Con Corporation, 2009.7, Rev-03
15. Dreyar, G., Koudounaris, A., Pratt, I.: The Reliability of Soldered or Epoxy Bonded Chip Capacitor Interconnections on Hybrids. IEEE Transaction on Parts, Hybrids and Packaging **13**(3), 218–224 (1977)
16. Application Note on RF Ceramic Chip Capacitors in High RF Power Applications, # 001-942 Rev. C; 4/05, American Technical Ceramics, USA
17. Application Note on ESR Losses in Ceramic Capacitors, # 001-923 Rev. D; 4/07, American Technical Ceramics, USA
18. Application Note on Advantages of Solid-State Relays over Electro-mechanical Relays, AN-145, dated 12/4/2012, IXYS Integrated Circuits Division, USA
19. Application Note 1036, Small Signal Solid State Relays, Avago Technologies, USA, 5965-5980EN-AUG 2010
20. Application Note 1046, Low On-Resistance Solid State Relays, Avago Technologies, USA, 5965-5980EN-AUG 2010
21. Application Note 5452, Design Considerations for Solid State Relays, Avago Technologies, USA, AV02-2321EN-MAR 2010
22. Application Note on Fuse Characteristics: Terms and Consideration Factors. Littelfuse Inc, USA (2009)

Chapter 4
Failure Mode and Effects Analysis

Abstract Failure Mode and Effects Analysis, popularly known as FMEA, is a technique that compels designers to think on the possible failures of equipment and their consequences for achieving total reliability. All engineers without exception do apply FMEA in the design of electronic equipment to prevent obvious design errors, and examples are provided for the application. However, the detection of latent design errors requires applying FMEA systematically. The terms, the benefits, and the application of FMEA are explained with examples from electronic designs for the benefit of circuit designers. FMEA improves the competence of designers and enhances the reliability and maintainability (R&M) of equipment. The documentation of FMEA becomes an asset to organizations.

4.1 Introduction

Failure Mode and Effects Analysis, popularly known as FMEA, is a technique that compels designers to think on possible failures for improving the reliability and maintainability of equipment. FMEA was developed by NASA for the design of aerospace and defense products [1, 2]. It is more popular with auto industries. FMEA is currently being applied by both auto majors and their ancillary units systemically across the world and the Society of Automobile Engineers (SAE) brought out a FMEA Standard for both design and manufacturing processes [1, 3]. Applying FMEA is explained for electronic designs benefiting circuit designers.

4.2 Significance of FMEA

Electronic equipment is an assembly of PCBs and modules. The Electrical performance failures in some PCBs and modules could result in equipment failure that violates statutory regulatory requirements. Inadvertent operational sequences

© Springer International Publishing Switzerland 2015
D. Natarajan, *Reliable Design of Electronic Equipment*,
DOI 10.1007/978-3-319-09111-2_4

during the operation and maintenance of equipment could cause serious failures. Disconnection of inputs to items might cause failures. The reliability techniques, de-rating for stress factors, and ensuring application reliability of components are not capable of preventing inadvertent, interconnection, and regulatory requirements related failures. Another reliability technique that focuses on the failures of items and their consequences is required for achieving the total reliability of equipment. The reliability technique is FMEA.

4.3 Examples of FMEA in Design

All engineers without exception do apply FMEA in the design of electronic equipment. Three examples are provided, supporting the application. Circuit designers select polarized PCB connectors for plug-in modules to prevent the inadvertent reversing of the modules in mounting. Selecting polarized PCB connectors is an example of FMEA.

In the design of cooling subsystem for VHF transmitters, both electronic and mechanical engineers are involved. The engineers discuss the method of providing interlock to prevent applying signal input to RF power amplifiers before switching on the cooling subsystem. The solution finalized in the discussion is an example of FMEA. The leader of a design team reviews the protection to RF Power Amplifier if the antenna is disconnected inadvertently during installation or maintenance. The solution finalized in the design review is an example of FMEA.

The three examples demonstrate preventive actions for obvious failure modes in the design of electronic equipment. However, the equipment could contain many latent failure modes. Systematic application of FMEA not only unearths the latent failure modes but also delivers more benefits to organizations and customers.

4.4 Benefits of FMEA

FMEA is brainstorming the circuit designs of electronic equipment by designers for possible failures before releasing provisional engineering drawings for developing the proto-models of the equipment. Designers experience a sense of accomplishment after performing FMEA and they become more competent with time. The benefits of FMEA are shown in Fig. 4.1.

The benefits are:

(i) Improving the reliability of design:
Identifying the existence of failures in design is fundamental for improving reliability. FMEA forces designers to think in the direction of identifying

Fig. 4.1 Benefits of FMEA

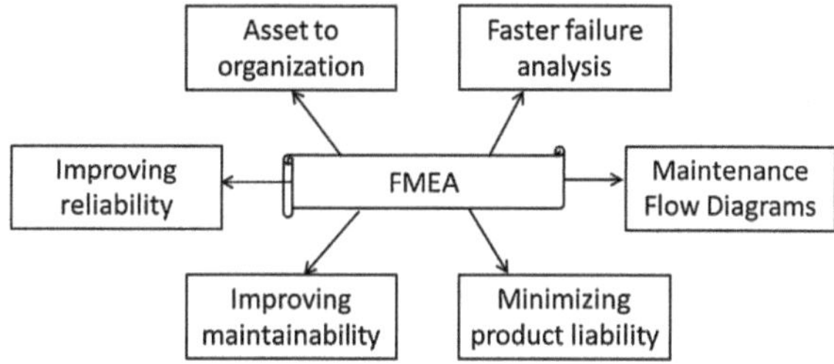

failures and their causes. Solutions are implemented to eliminate the causes of failures.

(ii) Improving the maintainability of design:
It might not be always feasible to eliminate the causes of failures. Fault detection and monitoring for failures are planned to improve the maintainability of design. Electrical parameters could be monitored for initiating preventive actions before the occurrence of failures. For example, the case temperature or collector supply current of high power RF transistor could be monitored.

(iii) Minimizing product liability claims [4]:
Failures might be observed in the operation of equipment at customer end. The cost of correcting field failures adds to product liability cost. The cost of correcting field failures is extremely high due to cascading effect. Re-designing of circuits, repeating design verification tests, changes in design documents, and retrofits are some of the high cost efforts to eliminate the causes of field failures. Preventing possible failures through FMEA is highly economical.

(iv) Maintenance flow diagrams:
Providing information for the troubleshooting of equipment in the form of flow charts is user-friendly. Maintenance Flow Diagrams (MFDs) are flowcharts linking the observed faults of equipment to defective modules for trouble shooting. They could be generated automatically from the records of FMEA.

(v) Faster analysis on functional failures:
The outputs FMEA could be used for fixing the electrical perfomance failures faster. Examples of functional failures are excessive harmonic levels, output power low, poor signal-to-noise ratio, etc.

(vi) Asset to organizations:
Schematic (circuit) diagrams of equipment are no doubt the most important output of design process. However, FMEA on modules provides insight into the design process with answers to the questions, how and why. The documentation of FMEA becomes an asset to organizations for correlating field performance with design and provides experience based inputs for new designs. The documentation of FMEA enables transfer of design to new engineers, when required.

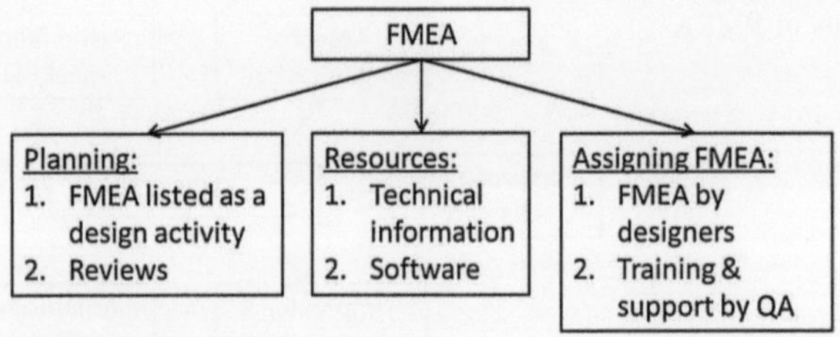

Fig. 4.2 Organizing FMEA in design

4.5 Organizing FMEA

De-rating and ensuring proper application for components are implemented by individual designers. However, the application of FMEA requires planning, software support, synergizing the expertise of designers, and reviews. The requirements for organizing FMEA for electronic equipment are shown in Fig. 4.2.

4.5.1 Planning in Design

The design and development of electronic equipment are planned after the approval of contract. Planning refers to the listing of the design and development activities with time schedules. FMEA should also be listed as one of the design activities. Reviewing the progress of FMEA should also be planned. The reviews demonstrate the commitment and support of management for administering FMEA in the design of equipment. Resources and the persons responsible for the activities should also be identified.

4.5.2 Resources

FMEA is a knowledge-based activity. Technical information and software support are essential for performing FMEA. Technical information is required for identifying possible failures in modules, analyzing circuits for fixing the causes of failures, and listing preventive actions to eliminate the causes of failure. The sources of technical information for applying FMEA are:

- Application notes of semiconductor devices
- Literature
- Reputed libraries across the world

4.5.2.1 Software Support

Administrative efforts are required for performing FMEA. Software support is essential for minimizing the administrative efforts so that designers could focus on the technical aspects of FMEA. Software is used for documenting the outputs of FMEA and for deriving the benefits of the analysis. Excel tool with macros is adequate for COTS items such as SMPS and Control units but, FMEA software is required for electronic equipment. Provisions are also required to record the changes in FMEA without deleting previously recorded information as the design of equipment progresses. More information for the design of FMEA software is available in Chap. 8.

4.5.3 Assigning FMEA

Electronic equipment contains PCBs, modules, and units. Understanding the functioning of the PCBs, modules, and units of electronic equipment is the core expertise required for applying FMEA. As the expertise is available with designers, performing FMEA should naturally be assigned to the designers. However, there are challenging roles for QA in FMEA. The roles of QA are:

(i)   Conducting training programs in FMEA for designers.
(ii) Documenting the failure modes and causes of electronic components. Identifying them is one-time activity and they are common to all electronic equipment.
(iii) Developing the library of preventive actions (Sect. 4.7.2).
(iv) Designing and developing FMEA software for designers (Sect. 8.1.2).

4.6 Terms in FMEA

It is convenient to document FMEA in the form of a spreadsheet for deriving the outputs of the analysis. The column headings of the spreadsheet are common to all the PCBs and modules of equipment. Most of the column headings are self-explanatory, and some headings are unique. The unique headings are the FMEA terms and they are:

(i)   Failure mode
(ii) Failure effect
(iii) Failure cause
(iv) Preventive actions

The hierarchical levels (indenture levels) of electronic equipment and their linkages should be documented for understanding the FMEA terms. The indenture

Fig. 4.3 Indenture levels of FM Transmitter

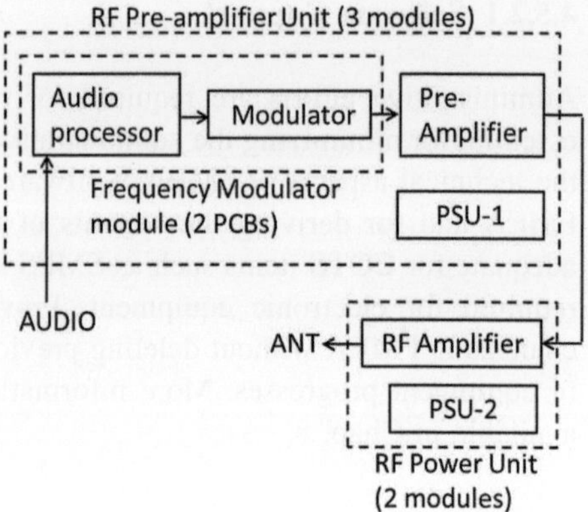

levels are explained for a simplified FM Transmitter, shown in Fig. 4.3. The transmitter is the top indenture level and it has three lower indenture levels. The lower levels are unit, module, and PCB. The units of the FM Transmitter are RF Pre-amplifier and RF Power. The RF Pre-amplifier unit has three modules such as Frequency Modulator, Pre-amplifier, and PSU-1. The Modulator module has two PCBs at the lowest indenture level and they are also shown in the figure. The RF Power unit has two modules such as RF Amplifier and PSU-2. FMEA terms are explained using the indenture levels of the FM Transmitter shown in Fig. 4.3.

4.6.1 Failure Mode

Failure mode of an electronic item is defined as the way or the manner in which the item fails. The item could be a component or a PCB or a module or a unit. One or more failure modes could exist for an item and they are assumed to have occurred for the purpose of FMEA. The failure modes of an item are identified by finding answers to the question, "How does the item fail?" [5].

Consider the Frequency Modulator module shown in Fig. 4.3. The failure modes of the module are identified by finding answers to the question, "How does the Frequency Modulator module fail?". Two failure modes of the modulator are identified and they are tabulated in Table 4.1. The failure modes of modules must be unambiguous and independent as explained in Sect. 4.6.5 for deriving the benefits of FMEA.

Table 4.1 Failure modes of frequency modulator

Sl. No.	Failure modes
1	Random variation of carrier frequency
2	Drift in carrier frequency

4.6.2 Failure Effect

Failure effect is the undesirable consequence when the failure mode of an item occurs. The functions of an item are impaired by the failure effects of the item. One or more failure effects could be associated with one failure mode. The failure effects of a failure mode are identified by finding answers to the question, "What happens if the failure mode were to occur?". Identifying failure effects is explained for one of the failure modes of the Frequency Modulator.

Consider the failure mode, "random variation of frequency" of the Frequency Modulator in Table 4.1. The failure effects of the failure mode are identified by finding answers to the question, "What happens if random variation of carrier frequency were to occur in the Frequency Modulator?". The failure effect on the Frequency Modulator is "distortion due to jittering". The effect is termed as Local Effect as it pertains to the Frequency Modulator. The local effect of Frequency Modulator module could propagate to higher indenture levels in equipment and hence, the failure effects at the higher indenture levels of the module should be examined.

Figure 4.3 shows that the next higher indenture level for the Frequency Modulator is RF Pre-amplifier unit. As the output of the Frequency Modulator is connected to the RF Amplifier in the unit, the same failure effect, "distortion due to jittering," is observed at the output of the RF Pre-amplifier unit and the failure effect is termed as Next Higher Level Effect. The failure effect observed at the output of the FM Transmitter is, "poor signal-to-noise ratio" and it is termed as End Effect. The local, next higher level, and end effects induced by the failure mode of the Frequency Modulator are shown in Fig. 4.4 and in Table 4.2.

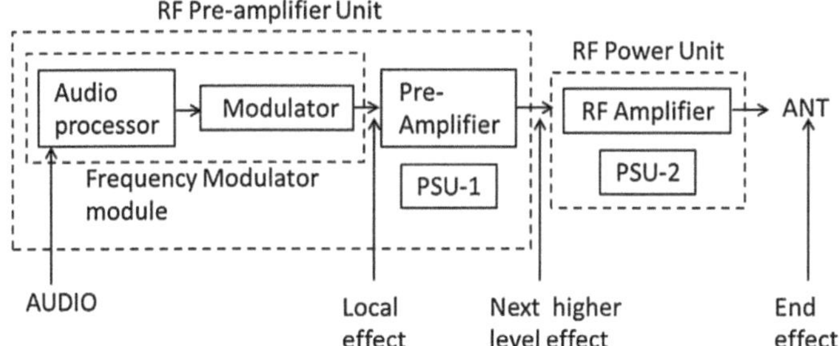

Fig. 4.4 Effects of module failure mode

Table 4.2 Failure mode of frequency modulator and its effects

Module	Failure mode	Failure effects		
		Local (frequency modulator module)	Next higher level (RF Pre-amplifier unit)	End (FM transmitter)
Frequency modulator	Random variation of carrier frequency	Distortion due to jittering	Distortion due to jittering	Poor signal-to-noise ratio

Next level and end effects might not exist for all items. For example, analyzing the failure modes of units of the FM Transmitter for failure effects, only the local and end effects need to be documented. For simple COTS items, local, next higher level, and end failure effects could mean the same.

4.6.3 Failure Cause

Failure cause analysis is the process of identifying the fundamental reasons for the occurrence of assumed failure modes of an item. The fundamental reasons are the root causes of the failure modes. One or more root causes could be associated with one failure mode. The failure causes of a failure mode are identified by finding answers to the question, "Why does the item fail?" [5].

Understanding the functioning of electronic modules is necessary for identifying the root causes of module failure modes. Consider the same failure mode of the Frequency Modulator in Table 4.1, i.e., "random variation of carrier frequency". Failure causes are identified by finding answers to the question, "Why does random variation of carrier frequency occur in Frequency Modulator?". The cause analysis indicates that there is no pre-emphasis circuit in the Frequency Modulator, causing the random variation of carrier frequency. The failure mode of Frequency Modulator and the cause of the failure mode are tabulated in Table 4.3.

4.6.4 Preventive Actions

Having identified failure modes, failure effects and the root causes of the failure modes of electronic modules, actions are initiated on the root causes to eliminate

Table 4.3 Failure mode of frequency modulator and the cause

Sl. No.	Failure mode	Failure cause
1	Random variation of carrier frequency	No pre-emphasis circuit in frequency modulator

Table 4.4 Failure mode of frequency modulator, cause, and preventive action

Sl. No.	Failure mode	Failure cause	Preventive action	
			Engineering reason	Action
1	Random variation of carrier frequency	No pre-emphasis circuit in Frequency modulator	The failure mode is significant at high modulation frequencies (e.g., music) and pre-emphasis circuit compensates the effects of the failure mode. As the FM transmitter transmits only voice signal, pre-emphasis circuit is not required	Nil

or mitigate the effects induced by the failure modes. The actions are preventive actions as the failure modes have not actually occurred and they are only postulated. The serious nature of local, next higher level, and end effects are examined for initiating preventive actions but the local effect provides the basis for deciding the preventive actions. Some criteria are required whether to initiate preventive actions or not.

The criteria, Risk Priority Number [4, 6 & 7] is used for initiating preventive actions and it is the product of ratings assigned to occurrence, severity, and detection in the scale from 1 to 10. Engineering reasoning is another criteria and it is more suitable for electronic equipment and COTS items to initiate preventive actions. The advantages of using engineering reasons as the criteria are:

(i) Engineering reasons are simple to understand as it speaks designer's language.
(ii) The engineering reasons could be readily reviewed by peers and team leaders for decision making.
(iii) The reasons could be correlated with observed field failures.
(iv) The reasons serve as references for future designs.

Failure mode, failure cause, and preventive action for the failure mode of the Frequency Modulator are indicated in Table 4.4 respectively. Engineering reason is indicated for the preventive actions.

4.6.5 Characterizing Failure Mode

Failure modes, identified for modules, should be unambiguous and independent. These are essential requirements of failure modes for deriving realistic maintenance flow diagrams, which are generated by software after completing FMEA on equipment. The requirements for characterizing the failure modes of modules are explained.

4.6.5.1 Unambiguous Failure Modes

Ambiguity might exist between the failure mode and local effect of an item and it causes the FMEA analyst to look for discriminators to improve fault diagnosis of system [5]. One method of identifying the unique failure modes is to consider the circuits of PCBs. A circuit is formed by one semiconductor device and its associated passive components. A PCB generally has many circuits with discrete devices and ICs. The unique failure modes of the PCB are identified from the failure modes of the circuits. If a PCB has one IC circuit, the unique failure modes are identified from the functional circuits of the IC.

4.6.5.2 Independent Failure Modes

Only independent failure modes of an item should be identified. The word "independent" has statistical significance. Consider a RF Power Amplifier, which is driven by a Pre-amplifier. Assume that the failure modes of RF Power Amplifier need to be identified. "No input to RF Amplifier from Pre-amplifier" is not a valid failure mode of the RF Power Amplifier as the failure mode is dependent on the failure of the Pre-amplifier. The failure modes of the RF Power Amplifier should be identified assuming that the modules supplying input signals and DC voltages to the Power Amplifier are functioning satisfactorily, and only such failure modes are independent.

4.7 FMEA at Module Level

Combining Tables 4.1, 4.2, 4.3, and 4.4 is an example of FMEA on the Frequency Modulator module and it is shown in Table 4.5. Additional columns as per the administrative needs of FMEA on modules could be added for the spread sheet. Module reference, next higher indenture level, date, etc. are examples of the administrative needs.

4.7.1 Interconnection FMEA

Input and output interconnections exist in electronic modules units and in the form DC and signal voltages. The interconnections could fail under vibration conditions or induced inadvertently during maintenance. The interconnection failures could induce soft or hard failure effects in items. "Output signal connector open" is an interconnection failure mode of Frequency Modulator module, and it produces soft failure effect, which is acceptable. Preventive actions are necessary for hard failures. Interconnection failures and their effects should be analyzed formally for

Table 4.5 FMEA on frequency modulator at module level

Failure mode	Failure effect			Failure cause	Preventive action	
	Local	Next level	End		Engineering reason	Action
Random variation of carrier frequency	Distortion due to jittering	Distortion due to jittering	Poor signal to noise ratio	No pre-emphasis circuit in frequency modulator	The failure mode is significant at high modulation frequencies. As the FM transmitter transmits only voice signal, pre-emphasis circuit is not required	Nil

all modules. Analyzing the effects of interconnection failure modes is a part of FMEA at module level, but it is presented separately to drive the concept of interconnection FMEA. The example of cooling system for FM Transmitter (Sect. 4.3) is used for illustrating interconnection FMEA, and it is presented in Table 4.6.

4.7.2 Library of Preventive Actions

Preventive actions for root causes to eliminate or mitigate the effects of failure modes do not automatically imply re-design of circuits. The actions could be selecting appropriate electronic components or adding protection and monitoring circuits or adding test points or a combination of them. Electronic components are available with built-in protection features. The list below indicates some of the preventive actions that could be considered for electronic equipment.

- Condition monitoring and preventing failures through software. For example, battery charging could be monitored to protect the battery from over-charging, which reduces the life of batteries due to heating.
- Adding protection circuits
- Front panel displays for monitoring
- Using pin or shell polarized connectors to prevent reverse-mating of plug-in modules
- Using different connector configurations to prevent the interchange of plug-in modules in mounting
- Using toggle switches with special actuators to prevent inadvertent switching operation
- Adding monitoring circuits
- Protection using fuses
- Interlocking of front panel operations
- Providing circuits or test points for fault detection

Table 4.6 Interconnection FMEA at module level

Failure mode	Failure effect			Failure cause	Preventive action	
	Local	Next level	End		Engineering reason	Action
Cooling system supply snaps	Cooling system does not function	RF power unit transistors fail	No transmitter output	Inter-connection failure	Protection circuit required to prevent the failure of RF power transistors	A sensing system monitoring the current drawn by the cooling system shuts down RF power unit if the current falls below the pre-set value

- Having alarm circuits
- Modifying circuit designs

4.7.3 Maintenance Flow Diagrams

End failure effects are the faults that could be observed in equipment. They are linked to defective units and modules by next higher level and local effects in FMEA. Software is designed to track the linkage and generate maintenance flow diagrams (MFDs) to isolate defective modules for the faults of equipment. One example of generating MFD from the FMEA documentation of Frequency Modulator (Sect. 4.7) is shown in Fig. 4.5.

Fig. 4.5 Maintenance flow diagram—an example

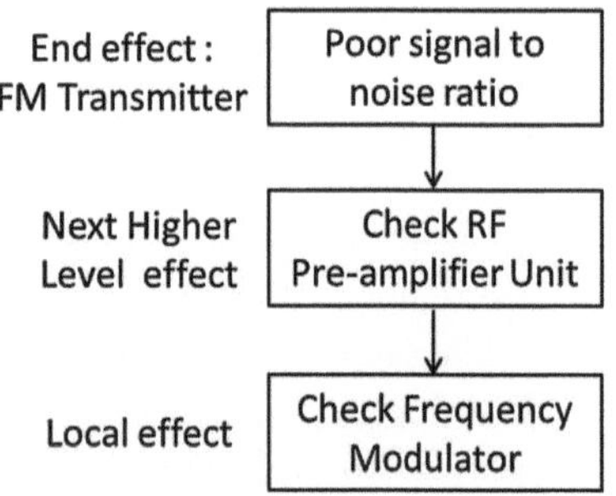

Table 4.7 FMEA at component level for RF filter

Failure mode	Failure effect			Failure cause	Preventive action	
	Local	Next level	End		Engineering reason	Action
Excessive change of cap. value	Pass band affected	Low signal band output	Low in-band power	General purpose capacitor used	Wrong selection of capacitor	Use temp. compensating type

4.8 FMEA at Component Level

Applying FMEA at component level requires higher level of efforts compared to FMEA at module level. Engineering expertise could be used to select components that have critical functional requirements for applying FMEA. For example, FMEA should be performed on the input and output coaxial connectors of microwave cavity band pass filter as the connectors experience mechanical stress combined with thermal stress in tuning operations. The replacement of failed connectors in completed microwave filters is an expensive task [6]. Expensive components should also be included for FMEA. Applying FMEA at component level is explained for the design of lumped (L-C) low pass RF Filter. The application of FMEA for one of the ceramic capacitor of the RF Filter is shown in Table 4.7. The failure effects, failure cause, and preventive actions for the failure mode of the capacitor i.e., "capacitance changes beyond acceptable limit" are documented.

The probability of occurrences of the failure modes of the ceramic capacitor should be considered to minimize efforts for the analysis. For example, additional failure modes are associated with the ceramic capacitor of the RF Filter and they are open circuit, short circuit, and have high dissipation factor. As the probabilities of occurrence of the failure modes are extremely low considering the voltage and frequency applications of the filter, FMEA need not be performed for the failure modes.

Assessment Exercises

1. Say True or False:

    (i) Proto-models of modules should be available for identifying the possible failure modes of the modules using FMEA.

    (ii) FMEA is a knowledge based activity to identify the possible failure modes of modules.

    (iii) Local, next higher level, and end effects of a failure mode could be identified for all electronic items.

    (iv) Failure cause is identified for the failure mode of an item and not for the failure effect.

    (v) The design actions to eliminate the failure causes identified in FMEA are corrective actions.

  (vi) After performing FMEA on items, initiating preventive actions is based on the severity of local effect or next higher level effect or end effect.

 (vii) Eliminating the failure causes identified in FMEA on modules is always accomplished by changing the circuit design of the modules.

(viii) The failure modes of a module are identified assuming the modules supplying input signals and DC voltages are functioning satisfactorily.

2. Define and explain the terms of FMEA:

   (i) Failure mode
  (ii) Failure effect
 (iii) Failure cause
 (iv) Preventive action

3. When performing FMEA on items, explain why the failure modes that are identified for the items should be unambiguous and independent.

Research Assignments

1. Perform FMEA on the products listed below, after identifying the indenture levels of the products. Identify at least one functionally critical component for the products and perform FMEA at component level also.

   (i) SMPS
  (ii) RF power amplifier
 (iii) LED Lighting unit
 (iv) Audio power amplifier
  (v) Battery charger

References

1. Montgomery, T.A., Pugh, D.R., Leedham, S.T., Twitchett, S.R.: FMEA automation for the complete design process. IEEE Proceedings of Annual Reliability and Maintainability Symposium, pp. 30–36 (1996)
2. Jordon, W.E.: Failure modes, effects and criticality analysis. IEEE Proceedings of Annual Reliability and Maintainability Symposium. pp. 30–37 (1972)
3. SAE International. Potential FMEA in Design (Design FMEA) and Potential FMEA in Manufacturing and Assembly processes (Process FMEA) Reference Manual. Technical Report J1739, 1994
4. Kara-Zaitri, C., Keller, A.Z., Barody, I., Fleming, P.V.: An improved FMEA methodology. IEEE Proceedings of Annual Reliability and Maintainability Symposium, pp. 248–252 (1991)
5. Spangler, C.S.: Equivalence relations within the failure mode and effects analysis. IEEE Proceeings of Annual Reliability and Maintainability Symposium. pp. 352–357 (1999)
6. Natarajan, Dhanasekharan: A Practical Design of Lumped Semi-Lumped and Microwave Cavity Filters. Springer, Berlin (2013)

Chapter 5
Overstress Analysis

Abstract Overstress is defined as exceeding the maximum ratings of components in electronic circuits. Although it is common practice to apply de-rating to components during design, electrical, and thermal overstress analyses are always performed to measure and confirm that components are operating within their maximum ratings. Vibration resonance overstress analysis is also conducted to identify fatigue-related mechanical design defects. Low-cost methods of conducting overstress analyses are presented in tutorial form.

5.1 Need for Overstress Analysis

Overstress is defined as exceeding the maximum rating of components in electronic circuits. De-rating and proper application of components prevent the application of overstress to components. Although it is common practice to apply de-rating to components during design, the portion of failures due to overstress is still very significant, approximately 25 % [1]. Hence, overstress analysis is always performed to measure and confirm that components are operating within their maximum ratings. The analysis includes mechanical overstress caused by vibration resonance.

5.2 Benefits of Overstress Analysis

Overstress analysis is a preventive action and the benefits of the analysis are shown in Fig. 5.1. The benefits are:

(i)   The failure of components is prevented.
(ii) Electronic components could still fail due to unforeseen reasons during manufacturing or usage. The failures of components could be fixed faster

© Springer International Publishing Switzerland 2015
D. Natarajan, *Reliable Design of Electronic Equipment*,
DOI 10.1007/978-3-319-09111-2_5

Fig. 5.1 Benefits of
overstress analysis

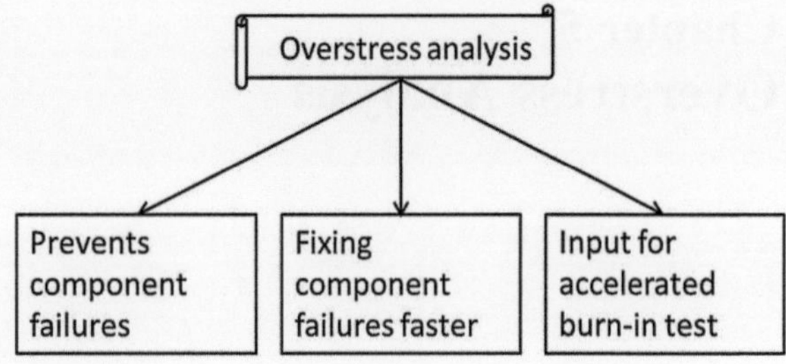

using the outputs of overstress analysis for initiating corrective actions.
Root cause analysis on component failures is explained in Chap. 7.

(iii)   The outputs of overstress analysis are used for deciding test temperature for
accelerated burn-in test. The method of designing the test is explained in
Chap. 6.

5.3 Types of Overstress

Electronic components could experience four types of overstress in their applications as shown in Fig. 5.2. The types of overstress are:

(i)   Transient electrical overstress
(ii)   Operating electrical overstress
(iii)   Thermal overstress
(iv)   Vibration resonance overstress.

5.3.1 Transient Electrical Overstress

High inrush current exists momentarily in motor and capacitive circuits during
power-on. Momentary high voltage is generated by inductive circuits during
power-off. Power-on and power-off conditions could occur in the normal functioning of equipment or when the plug-in modules of equipment are removed and
replaced during maintenance with equipment powered on. The momentary overstress is termed as transient electrical overstress. Components in the series path of
transient current and those in the shunt path of transient voltage might be
overstressed.

Transient electrical overstress is explained for a capacitive circuit. Capacitors
act as short circuit when equipment is powered on and they result in high inrush
current. The power-on transient current waveform is shown in Fig. 5.3. The peak
value of current is limited by the resistance of circuit and the duration of the

Fig. 5.2 Types of overstress

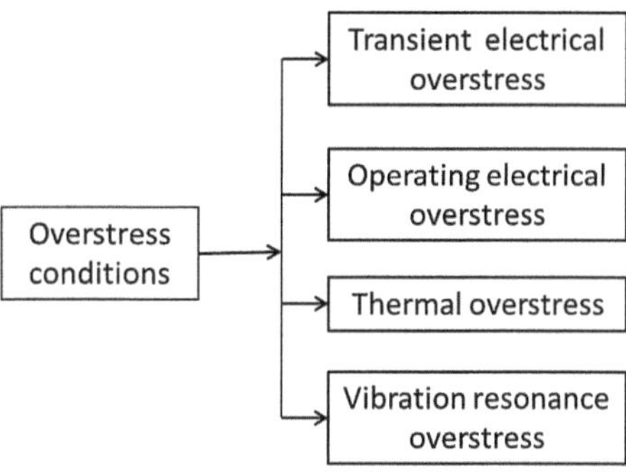

Fig. 5.3 Power-on transient
current in model

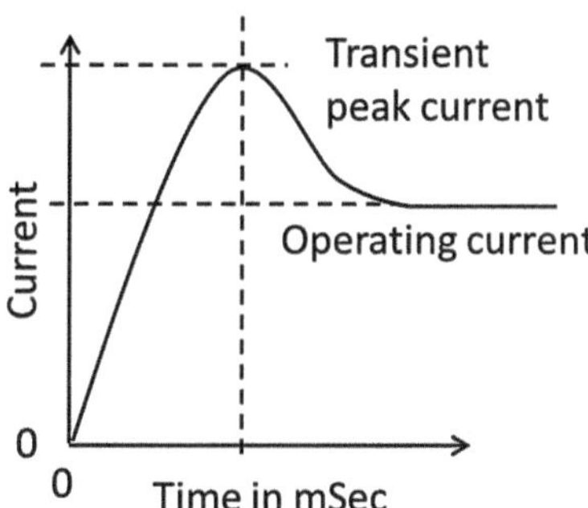

transient waveform is related to the time constant of the circuit. Waveforms with higher peak value and duration have higher transient energy.

5.3.2 Operating Electrical Overstress

Operating electrical overstress refers to the continuous or pulsed application of voltage or current or power exceeding the ratings of components in the functioning of electronic equipment. It is primarily caused by design errors in the estimation of stresses on components. Selecting 25 V aluminum electrolytic capacitor for a filter circuit at 28 V DC is an example of operating electrical overstress. Operating power overstress also causes thermal overstress.

5.3.3 Thermal Overstress

Like operating electrical overstress, thermal overstress on components is caused by design errors in the estimation of stresses. Thermal overstress occurs when ambient or surface or case or junction temperature exceeds the maximum ratings

Fig. 5.4 Sources of thermal
overstress

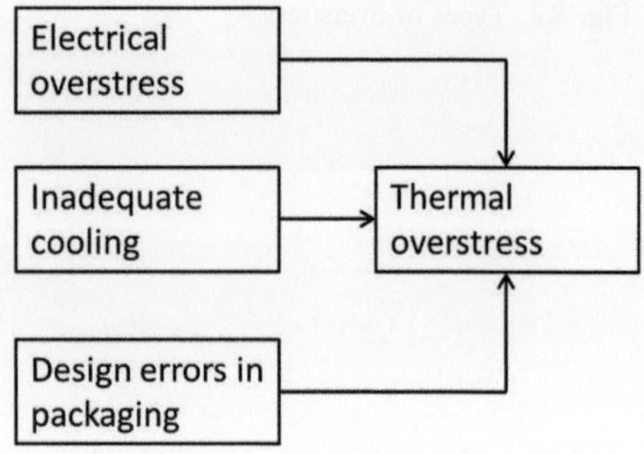

of components. The sources of thermal overstress are shown in Fig. 5.4. Thermal
overstress is caused by:

(i)   Excessive power dissipation or current increases operating junction or
      surface temperature exceeding the maximum rating of components.
(ii) Inadequate design in the methods of cooling results in excessive internal
      ambient temperature causing thermal overstress for components.
(iii) Design errors in packaging electronic components in PCBs could cause
      thermal overstress. A small signal transistor mounted near a heat dissipating
      wire wound resistor in a PCB experiences higher ambient temperature,
      which could result in operating junction temperature exceeding the absolute
      maximum rating of the transistor.

5.3.4 Vibration Resonance Overstress

Unlike electrical and thermal overstress, vibration resonance overstress is not
related to the maximum rating of components. Vibration resonance overstress
occurs due to improper mechanical mounting of components on PCBs or improper
mechanical design for the mounting of assembled PCBs on chassis. For example, it
is proper to mount a ½ W metal film resistor with axial leads by bending the leads
at 3 mm from their point of emergence and soldering them on PCB. However, it is
improper to mount a 12 W wire wound resistor in the same manner. The body of
the wire wound resistor should be rigidly clamped onto PCB by mechanical means
in addition to soldering as the resistor is heavier. Without the clamping, the wire
wound resistor mechanically vibrates between the soldered ends during the
transportation of equipment. The amplitude of vibration is highest at the
mechanical resonant frequency decided by PCB and other mechanical designs.
The vibration resonance causes overstress to solder joints and components,
degrading reliability. Solder joints are meant for electrical connections and they
are not capable of withstanding mechanical stresses.

5.4 Low Cost Methods for Overstress Analyses

Advanced and sophisticated techniques are available for thermal overstress analysis. Computational Fluid Dynamics (CFD) and Finite Element Analysis (FEA)-based software packages are used for predicting and measuring the operational temperature of PCBs, modules, and equipment to estimate the levels of overstress [2, 3]. Infrared-based thermal imaging equipment is also available for temperature measurements. Low-cost methods for conducting overstress analyses on electronic equipment are also available and they are presented. The presentation is based on applying the methods on a wide range of electronic modules and equipment.

5.4.1 Planning

The proto-models of PCBs, modules, units, and equipment should be verified for compliance to electrical performance specifications prior to conducting overstress analyses. Major circuit design changes, if effected after overstress analysis, would require repetition of the analysis. Overstress analyses are always conducted at ambient temperature.

5.4.1.1 Items for Overstress Analyses

Overstress analyses could be conducted on finished COTS items as the size and weight of the items are not large. However, conducting overstress analyses on finished larger electronic equipment like FM Transmitter is not recommended as the focus for identifying overstressed components might be lost. Hence, the analyses are conducted at lower indenture levels of equipment by grouping PCBs and modules. Grouping of PCBs and modules is decided by the engineering considerations, listed below:

(i)   Functional group of modules that generate worst level of transient electrical overstress and thermal overstress should be identified.

(ii) The test item for electrical and thermal overstress analyses should be independently testable with standard test equipment.

(iii) Efforts for overstress analyses should be minimized.

(iv) The weight of the test item should be limited from 3 to 5 kgs so that general purpose vibration fixtures could be used for vibration resonance overstress analysis.

5.4.1.2 Exclusions for Overstress Analyses

The PCBs and modules of equipment could be excluded from electrical and thermal overstress analyses, provided the operating conditions of the items are benign. The benign operating conditions are defined as:

(i) The worst-case internal operating temperature of items does not exceed 60 °C.
(ii) There are no thermally stressed components.
(iii) Sources that generate transient voltage or current do not exist.
(iv) The item does not experience transient electrical overstress or thermal overstress induced by other items in equipment.
(v) Operating voltages are within limits as verified from the list of components.

Small rigidly mounted PCBs with surface mount components could be excluded from vibration resonance overstress analysis. PCBs of size 100 mm × 100 mm could be considered safe for exclusion. All the exclusions should be reviewed and approved to prevent inadvertent design errors.

5.4.2 Resources

The general purpose test equipment and accessories for electrical and thermal overstress analyses are listed below. Additional equipment and accessories might be required for conducting electrical overstress analysis. Input signal sources are required for amplifier modules. Mains operated modules require attenuating probes for oscilloscope. RF modules require 50 Ω standard termination or RF power meter. The general purpose test resources are:

(i) DC Power Supplies
(ii) Digital storage oscilloscope, 10 MHz or better
(iii) Calibrated shunts, 1, 5 mΩ, etc., as required
(iv) Clip-on DC Ammeter with output port for oscilloscope
(v) Digital thermometer with thermocouple probes
(vi) Temperature indicating stickers, if required
(vii) General purpose digital multi-meter.

Facilities for vibration resonance overstress analysis are expensive and the facilities in electronics test centers could be utilized. The facilities for vibration resonance overstress analysis are:

(i) Electro-dynamic sinusoidal vibration equipment
(ii) Resonance free test fixture for mounting item under test
(iii) Video stroboscope for resonance search
(iv) Provisions to measure vibration resonance amplitudes.

5.5 Measuring Transient Electrical Overstress

Measuring transient electrical overstress should be emphasized for electronic modules as most of the design-related component failures are caused by the overstress. Transient currents and voltages should preferably be measured without disturbing the electrical functioning of the item. Normal electrical operating conditions (DC supply and signal inputs) are set up for the item under overstress analysis. The output of the item is suitably terminated. The general procedure for measuring transient electrical overstress is shown in Fig. 5.5. Application notes of semiconductor devices should be referred for measuring transient electrical overstress such as dv/dt on MOSFETs with inductive loads [4].

5.5.1 Modules for Measuring Overstress

In general, modules having high values of reactive components and those operating in transient electrical stress environment should be analyzed for overstress. DC power supplies are examples of modules having high values of inductors and capacitors. Electronic modules for automobiles should be analyzed for overstress as transient electrical stress environment exists in automobiles. Modules having excessive transient electric current could also be identified by capturing power-on current waveform of the modules using oscilloscope as explained in Sect. 5.5.2. For example, if a module measures peak current of 3.7A and normal operating current of 1.1A as shown in Fig. 5.6, the module should be analyzed for transient current overstress.

In some circuits, transient current could result in generation of transient voltage. For example, assume an inductor-capacitor ripple filter circuit with capacitor switched on in the application. When the capacitor is switched on, there is high inrush current through the inductor. Once the voltage across the capacitor reaches full voltage, the energy stored in the inductor raises the voltage across the capacitor momentarily [5]. Expertise in circuit design and understanding components add value to transient overstress analysis.

5.5.2 Measuring Transient Current

Assume that a module contains a capacitive load circuit as shown in Fig. 5.7. 12 V DC power supply is connected to the circuit. The regulated power supply should be capable of maintaining the set voltage at least up to twice the expected peak value of inrush current. A bounce-free switch and a calibrated shunt are also shown in the circuit for measuring power-on inrush current of the module. Modern

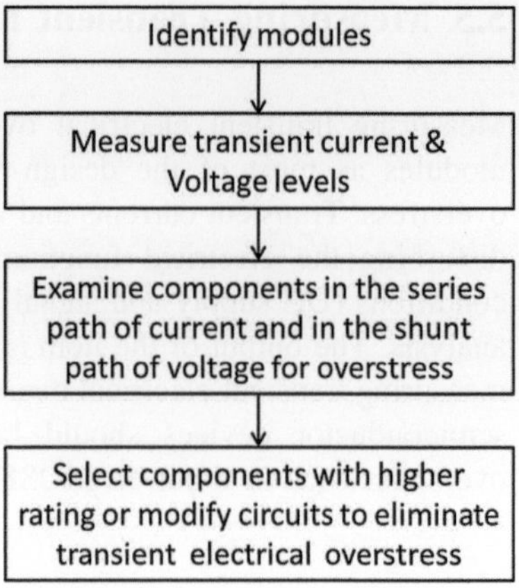

Fig. 5.5 Transient electrical overstress analysis procedure

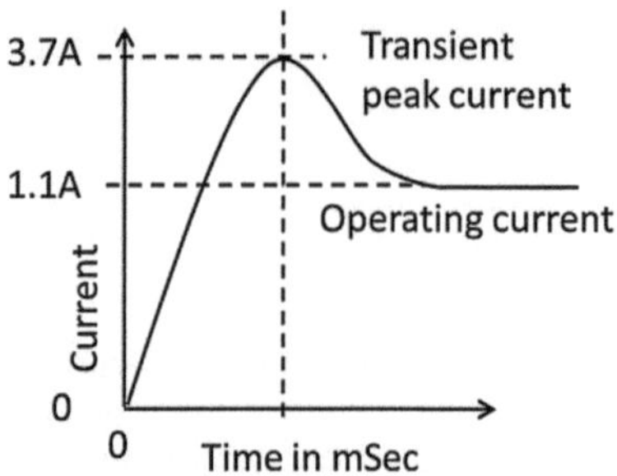

Fig. 5.6 Power-on transient current in a module— Example

high-performance power supplies are available for characterizing inrush current and they eliminate the need for bounce-free switch and the calibrated shunt [6].

The voltages before and after the shunt are measured by the digital storage oscilloscope. The current drawn by the module is computed from the difference between the voltages and the resistance value of the shunt. Alternatively, the current drawn could be measured using Clip-on DC Ammeter having acceptable transient response and scope. The voltage scale in the scope could be re-calibrated to read current directly by using the resistance value of the shunt or by using the ratio output of the clip-on ammeter.

The circuit contains 560 μF capacitor and causes high inrush current when the module is powered on. Before closing the switch to simulate power-on condition for the module, the capacitor is discharged for residual charges to avoid measuring erroneous value of inrush current. Triggering conditions are adjusted in Digital Storage Oscilloscope to capture the one shot power-on transient input current. The switch is closed capturing the transient current waveform. The peak value of transient (inrush) current and duration are read in the scope.

The components in the series path of the inrush current are the transistor, TR_1, and the resistor, R_1. Application notes of the components are referred for computing the acceptable level of transient current. If adequate information is not

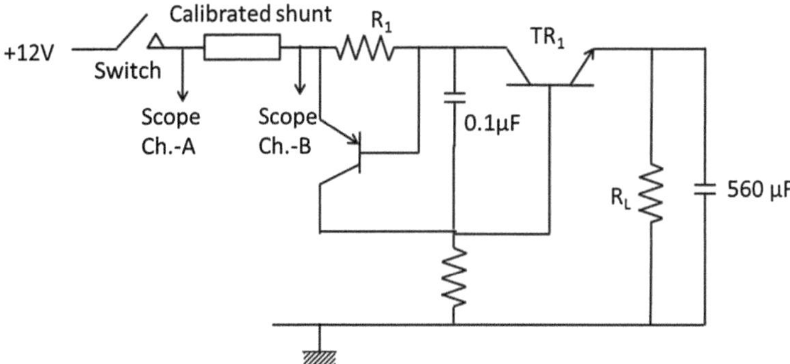

Fig. 5.7 Measuring transient current

available, such circuits are identified and evaluated for reliability by conducting usage on–off stressing as explained in Sect. 6.7.2, Chap. 6. Components with higher rating are selected or design changes are effected to eliminate transient current overstress.

5.5.3 Measuring Transient Voltage

The procedure for measuring transient voltage is the same as that for measuring transient current except that the generated voltage by inductive loads is measured under power-off condition. Power inductors, electro-mechanical relays, and solenoids are examples of inductive loads. The oscilloscope captures the power-off transient voltage waveform. The components in the shunt path of the transient voltage are identified by studying the circuit of the item under test. The components that are not capable of withstanding the measured transient energy (peak value of voltage and duration) are electrically overstressed. Components with higher rating are selected or design changes are effected to eliminate transient voltage overstress.

5.6 Measuring Operating Electrical Overstress

Operating electrical overstress analysis is integrated with transient electrical overstress analysis as the test setup for both the analyses is identical. The switch in Fig. 5.7 is closed for measuring electrical overstress on components. The schematic diagram of the item under overstress analysis is examined and the operating stress levels (voltage, current, and power dissipation) on components are computed. If required, the operating electrical stress levels are measured using the oscilloscope for confirming the computations. The operating stress levels are compared with the ratings of the components to identify electrically overstressed components except for power overstress. The power dissipation capability of components is linked to the operating temperature of components (see Sect. 2.6.2.1). The power overstressed components are identified after measuring worst-case operating temperature for the components in thermal overstress analysis. Components with higher ratings are selected or design changes are effected to eliminate operating electrical overstress.

5.6.1 Practical Considerations

The ratings of passive components are high compared to the application needs of most electronic circuits in electronic equipment. Operating electrical overstress analysis could be limited to thermally stressed components. Hence, the schematic diagram of item under analysis is examined to identify the components that are thermally stressed and the operating electrical stress levels on the components are estimated or measured.

5.7 Measuring Thermal Overstress

Thermal overstress analysis is conducted in two stages as thermal stress analysis and thermal interaction analysis. The stages for measuring thermal overstress are shown in Fig. 5.8. Thermal stress analysis is conducted on the independently testable items (group of PCBs and modules) of equipment. Thermal interaction stress analysis is conducted on assembled equipment after completing thermal stress analysis on the independently testable items.

In thermal stress analysis, thermally stressed components in the items (group of PCBs and modules) of equipment are identified and the case or the surface temperatures of the components are measured. The worst-case operating temperatures of the components and the level of power stress are computed after conducting thermal interaction stress analysis. Thermal stress analysis and thermal interaction analysis are explained with examples.

5.7.1 Thermal Stress Analysis

The general procedure for measuring thermal stress is shown in Fig. 5.9. Thermal stress analysis is conducted at ambient temperature and it is explained for an item containing PCBs in a metallic enclosure. Let the item be denoted as Module-A. The module is placed away from air draught through windows or fans, creating still

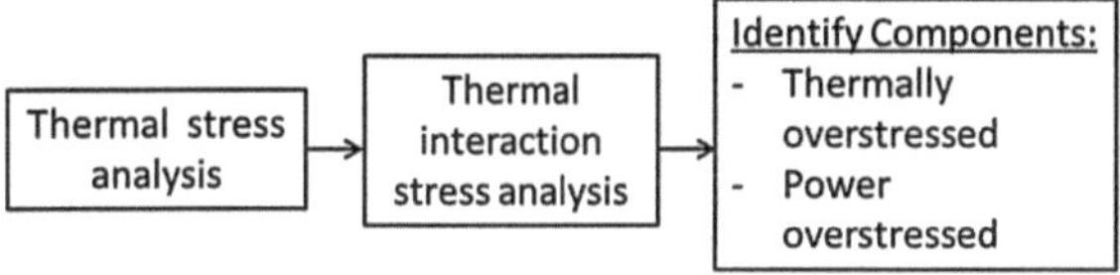

Fig. 5.8 Stages of thermal overstress analysis

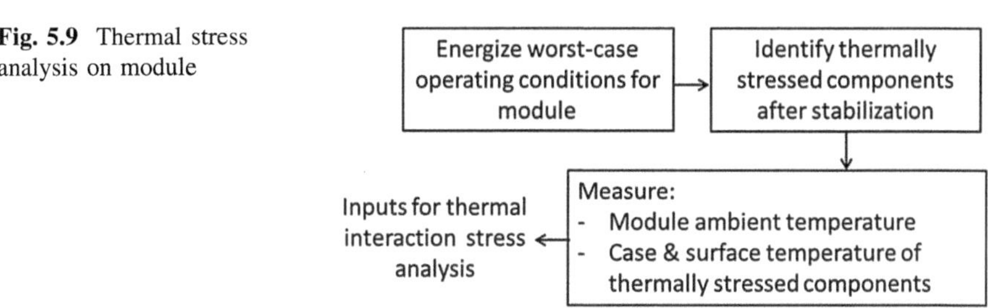

Fig. 5.9 Thermal stress analysis on module

air conditions for conducting thermal stress analysis. Arrangements for DC supply voltage, applying appropriate input signal and terminating the output for the module are organized.

5.7.1.1 Identifying Thermally Stressed Components

Digital thermometer with thermocouple probes is used for thermal stress analysis. The thermocouple thermometer is acceptable and accurate. PCBs are probed with thermocouple to verify that the temperatures given by infrared scan are accurate [7]. The cover of the Module-A is removed exposing the PCBs for measuring the operating temperatures of components. A thin flexible wire type thermocouple probe is fixed using a small adhesive tape on the wall of module enclosure. The location on the wall is selected such that it represents the ambient temperature of the module. Worst-case electrical operating conditions are set up for the Module-A and the conditions are maintained for a minimum period of 2 h to ensure thermal stabilization.

The thermocouple on the wall of the module enclosure indicates the operating ambient temperature of the module. Components are probed by a pencil type thermocouple probe. The components that have operating surface or case temperatures 5–10 °C higher than the ambient temperature are thermally stressed. When probing the case temperature of RF transistors, the temperature reading on the Digital thermometer might fluctuate due to RF interference. The input RF signal is disconnected momentarily for reading the case temperature of the transistors. The thermally stressed components are identified for measuring case and surface temperatures.

5.7.1.2 Measuring Temperature of Thermally Stressed Components

Measuring the case temperature is explained for one of the thermally stressed transistors (TR$_1$) of Module-A. The power dissipation and characteristics of transistor TR$_1$ are assumed as:

Power dissipation (P_D) = 11 W
Power dissipation at 25 °C case temperature = 30 W max.
Thermal resistance, (θ_{j-c}) = 4.2 °C/W
Maximum junction temperature $(T_j \text{ max})$ = 150 °C

The thermocouple fixed on the wall of the module is not disturbed. Thin flexible wire type thermocouples are fixed to the case of the transistor TR$_1$ of Module-A. Additional components could be included for temperature measurement based on the engineering analysis of the schematic diagram of Module-A. The cover of Module-A is fixed and the thermal stress analysis is repeated as described above. The temperatures displayed by the thermocouple probes after thermal stabilization of the module are recorded and are shown in Table 5.1. The computed operating junction temperature of the transistor TR$_1$ is also shown in Table 5.1. The worst-case operating junction temperature of the transistor is computed after conducting thermal interaction stress analysis.

5.7.2 Thermal Interaction Stress Analysis

Thermal interaction stress analysis is conducted at ambient temperature on finished equipment after completing thermal stress analysis on all the independently testable items (group of PCBs and modules) of the equipment. The general procedure for measuring thermal interaction stress on equipment is shown in Fig. 5.10.

The assumptions for thermal interaction analysis are:

- The equipment is FM Transmitter.
- The maximum operating ambient temperature $(T_{A\text{-}TR\text{-}max})$ of the FM Transmitter is 45 °C.
- Module-A is part of the FM Transmitter.
- Thermal interaction stress analysis on the transmitter is conducted at an ambient temperature $(T_{A\text{-}TR})$ of 26 °C.

Table 5.1 Operating conditions of the transistor T_{R1} in thermal stress analysis

Parameter	Temperature (°C)
T_{A1}, operating ambient temperature of Module-A	28
Measured case temperature (T_{C1}) of transistor, T_{R1}	80
Operating junction temperature (T_{j1}) of TR$_1$ = $[T_{C1} + \theta_{j-c} * P_D]$	126

Fig. 5.10 Thermal
interaction stress analysis
on equipment

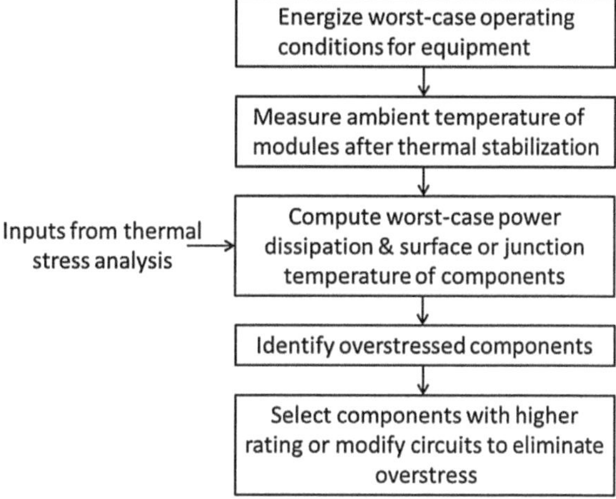

The transmitter is placed away from air draught through windows or fans, creating still air ambient conditions for stress analysis. Worst-case electrical operating conditions are set up for the FM Transmitter. The power dissipation of the FM Transmitter and the thermal interaction between the modules of the transmitter increase the internal ambient temperatures of modules. The internal ambient temperatures of the modules are measured to identify thermally and power overstressed components.

5.7.2.1 Measuring Internal Ambient Temperature of Modules

Measuring the internal ambient temperature of Module-A is explained. Thin flexible wire type thermocouple probe is fixed onto the wall of Module-A and other items (modules and units) of the FM Transmitter. The locations of fixing the probes should preferably be the same as those used for thermal stress analysis on the items. If required, additional PCBs and critical semiconductor devices are also included for temperature measurements. The FM Transmitter is powered-on for sufficient duration so that the transmitter could attain thermal stabilization. Referring to military standards, the transmitter is considered to have attained thermal stabilization if two consecutive temperature readings recorded at 15 min interval do not differ by more than 0.5 °C. The temperatures indicated by the thermocouple probes are recorded. It is assumed that the thermocouple probe on the wall of Module-A has recorded 37 °C.

The internal ambient temperature of the Module-A has increased from 28 °C (T_{A1}) in thermal stress analysis to 37 °C (T_{A2}) in thermal interaction analysis on the FM Transmitter. The increase in temperature is 9 °C. The operating case and junction temperatures of transistor TR_1 of Module-A in thermal interaction stress analysis would have also increased by 9 °C, ignoring thermal-related errors in the computation. The recalculated temperatures are shown in Table 5.2.

5.7.2.2 Computing Worst-Case Operating Temperature of Components

Components experience the worst-case operating temperatures when the FM Transmitter operates at the maximum operating ambient temperature. If the operating ambient temperature of the FM Transmitter is increased from 26 °C ($T_{A\text{-}TR}$) to 45 °C ($T_{A\text{-}TR\text{-}max}$), the internal operating temperatures of modules and components would also increase proportionately by 19 °C (45–26 °C). The computed worst-case operating conditions of Module-A and transistor TR_1 are shown in Table 5.3. The estimated worst-case operating temperatures would be marginally less than the computed values due to natural convection air cooling. Accuracy in the computation of worst-case temperatures is not required for applying de-rating or for selecting better components.

The analysis shows that the worst-case operating junction temperature of the transistor TR_1 exceeds the maximum junction temperature and is caused by power overstress on the transistor. The power rating of the transistor TR1 at case temperature 108 °C is 10 W (see Sect. 2.6.2.1), whereas the transistor is dissipating 11 W. Selecting a transistor with higher rating or improving heat transfer or

Table 5.2 Temperatures in thermal interaction analysis (FM TX at 26 °C)

Parameter	Temperature (°C)
$T_{A\text{-}TR}$, Operating ambient temperature of FM Transmitter	26
T_{A2}, Internal operating ambient temperature of Module-A	37
T_{C2}, Case temperature of transistor, T_{R1}	89
T_{j2}, Junction temperature of $TR_1 = [T_{j1} + T_{A2} - T_{A1}]$	135

Table 5.3 Worst-case temperatures at FM Transmitter at 45 °C

Parameter	Temperature
$T_{A\text{-}TR\text{-}max}$, Maximum operating ambient temperature of FM Transmitter	45 (°C)
$T_{A\text{-}wc}$, Worst-case internal operating ambient temperature of Module-A $= [T_{A2} + (T_{A\text{-}TR\text{-}max} - T_{A\text{-}TR})]$	56 (°C)
T_{c3}, Worst-case operating case temperature of $TR_1 = [T_{C2} + (T_{A\text{-}TR\text{-}max} - T_{A\text{-}TR})]$	108 (°C)
$T_{j\text{-}wc}$, Worst-case operating junction temperature of $TR_1 = [T_{j2} + (T_{A\text{-}TR\text{-}max} - T_{A\text{-}TR})]$	154 (°C)
T_j max, Maximum junction temperature of TR_1	150 (°C)
Power dissipation of transistor, TR_1, at $T_{C3} = 108$ °C	11 W
Power rating of transistor, TR_1, at $T_{C3} = 108$ °C (see Sect. 2.6.2.1)	10 W max.

initiating design changes should be considered to limit the operating junction temperature of the transistor TR_1 to 135 °C.

5.8 Vibration Resonance Overstress Analysis

Environmental requirements are part of the overall technical requirements of equipment and they are specified by customers. The environmental requirements include temperature range, humidity, vibration, shock, and other requirements such as altitude, acceleration, and salt fog. Equipment experience vibration and shock during transportation to the destinations of customers and during operation in vehicle mounted equipment. Most electronic components are qualified for shock and vibration. Although they contribute significantly for the reliability of equipment, vibration design is necessary to satisfy the shock and vibration requirements of equipment.

5.8.1 Vibration Fatigue-Related Defects

The vibration design of equipment involves deciding the materials of hardware, the methods of mounting the electronic and mechanical hardware, selection of resilient mounts, etc. The design book, Vibration Analysis for Electronic Equipment [8] could be referred for details. One of the requirements of vibration design is to push up the vibration resonance frequencies of hardware to eliminate or minimize the fatigue-related failures of electronic components in equipment. The fatigue-related mechanical design errors that could be detected by vibration resonance overstress analysis are:

   (i)   Design of PCBs (thickness and size)
   (ii)   Inadequacies in mounting of PCBs
   (iii) Improper mounting of heavier electronic components
   (iv)   Inadequate clamping for cable bunches
   (v)   Inadequacies in the design and mounting of mechanical hardware.

The general procedure for conducting vibration resonance overstress analysis is shown in Fig. 5.11.

5.8.2 Vibration Fixture

Rarely, item under test is mounted directly on the vibration table. Vibration fixture is used for mounting the item under test as it facilitates conducting vibration test in the three mutually perpendicular planes. Vibration fixture is designed with minimum mass and maximum stiffness to ensure the vibration resonant frequencies of the

Fig. 5.11 Vibration resonance overstress analysis procedure

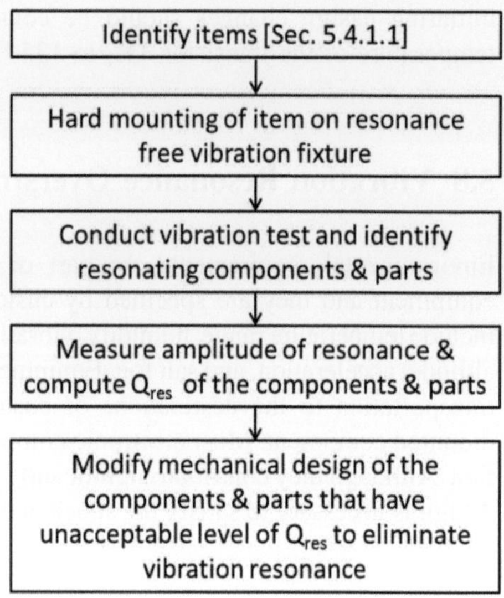

fixture are outside the frequency band of interest. The fabricated vibration fixture is evaluated by mounting accelerometers at different locations on the fixture to confirm that there are no resonant frequencies in the frequency band. The fixture is evaluated in the three mutual perpendicular mounting orientations. Vibration fixture is designed so that it could be mounted on the shaker and on the slip table of vibration machine.

5.8.3 Test Conditions

Standards for environmental test methods specify both random vibration and sinusoidal vibration. Random vibration is specified for precipitating interconnection defects and for qualifying equipment. Detailed information regarding random vibration is provided in Chap. 6. Sinusoidal vibration testing is appropriate for identifying vibration resonances in the frequency band of equipment [9, 10]. The variables defining the test conditions for the sinusoidal vibration resonance overstress analysis are:

   (i)   Sweep frequency band
  (ii)   Amplitude
 (iii)   Frequency sweeping.

5.8.3.1 Sweep Frequency Band

Unlike the vibration test for qualifying equipment, the test conditions for vibration resonance overstress analysis are flexible [9, 10]. The sweep frequency band for the sinusoidal vibration testing is selected considering the following information:

(i)   Resonances at low frequencies cause higher level of fatigue-related damages to hardware than the resonances at high frequencies. Customers indicate minimum vibration resonance frequency (Ex.: 100 Hz min.) for the electronic units installed in critical systems as part of operating environmental requirements and it means that no vibration resonances are permitted below 100 Hz in hardware.

(ii) The frequency band for rail and road transportation could be used for deciding the frequency band for the vibration resonance test. IEC 61373, Railway Applications—Rolling Sock Equipment—Shock and Vibration Tests, specifies (5–150) Hz. Telecom Standards specify (10–350) Hz. Military standards specify (5–500) Hz.

Analyzing electronic hardware for vibration resonances in the frequency band (10–350) Hz or lower would be adequate for most electronic equipment and COTS items. Industrial practices vary in deciding the frequency band for the vibration test. Vibration resonances above 150 Hz are less damaging to hardware as the amplitude of vibration is low.

5.8.3.2 Amplitude

Peak values are specified for the amplitude of sinusoidal vibration. For example, the Standard for environmental test methods on components, MIL-STD-202G, specifies the amplitude of sinusoidal vibration as $\pm0.06''$ (1.52 mm) double amplitude or 10 g peak, whichever is less in the frequency band (10–500) Hz, with crossover frequency at 55 Hz. The frequency at which the vibration amplitude computed from the gravity unit equals the specified peak value of double amplitude in inches is defined as the crossover frequency. Peak amplitude at any frequency could be obtained from the expression

$$D = \frac{386.086 * g_{\text{peak}}}{2 * \Pi^2 * f^2}$$

D Peak double amplitude in inches
g_{peak} Peak value of acceleration
f Frequency in Hz

The amplitude of vibration specified in the Standards for environmental test methods is not relevant for vibration resonance testing. The amplitude of vibration is maintained low to prevent possible damages to electronic and mechanical hardware during testing [10]. With low forcing amplitude, the jump phenomenon is absent and the system frequency response is closer to linear [9]. The absence of the jump phenomenon prevents the missing of resonant frequencies in the frequency band of item under test. A low level of 0.25 g sinusoidal sweep was performed to determine the resonant frequency of test PCB [11]. Very low amplitude level poses difficulties for measuring the resonant amplitudes of parts. Amplitude level up to 1 g peak but not exceeding 0.006 inches (0.15 mm) peak double amplitude could be considered for conducting sinusoidal vibration test for identifying resonant frequencies. The amplitude level could be reduced considering the capabilities of accessories for detecting and measuring vibration resonances.

5.8.3.3 Frequency Sweeping

The rate of sweeping is stipulated for sinusoidal vibration test in Standards such as one octave per minute. Such sweep rates are high for vibration resonance test as the detection of resonant frequencies is difficult. Any convenient slower rate of sweeping could be decided. Low sweep rate and sufficient dwell time at the resonant frequencies for noting the response are recommended [9]. The number of sweeps in each of the three mutually perpendicular direction of vibration is based on the successful detection of resonant frequencies in the sweep frequency band.

5.8.4 Defining Q-Resonance

Item under test is mounted on a vibration machine using a fixture for conducting vibration resonance analysis by frequency sweeping. The shaker of the vibration machine is excited using high power amplifier units generating sinusoidal vibrations in the shaker of the machine. The peak amplitude at which the shaker is excited is the drive amplitude of vibration. At any frequency of vibration, if the amplitude of vibration of parts is the same as the drive amplitude, the parts are not experiencing vibration resonance. If the amplitude of vibration of parts is higher than the drive amplitude, the parts are experiencing vibration resonance, i.e., there is amplification of amplitude at resonant frequencies. Q-Resonance (Q_{Res}), the amplification factor at resonant frequencies, is defined as

$$Q_{Res} = \frac{\text{Vibration resonant amplitude of part}}{\text{Drive amplitude}}$$

5.8.5 Identifying Stressed Parts

Item under test is preferably hard-mounted on vibration fixture that is mounted on the platform of vibration machine. The covers of the item are removed so that all the PCBs are visible for vibration resonance search. The test conditions for vibration are set in the vibration machine. Stroboscope with neon lamp or video-stroboscope is used for detecting vibration resonance on the parts of the item under test by scanning the parts. The stroboscope is synchronous with the sweep frequency of vibration.

The item is subjected to the sinusoidal sweep vibration test. Both the mechanical parts and the electronic components, including the PCB of the item, are scanned by the stroboscope. The parts and components that appear stationary are not resonating and those that appear oscillating are resonating. The resonating parts are identified and the frequencies of resonance are recorded.

5.8.5.1 Measuring Q-Resonance

Noncontact laser method is used for measuring the resonant amplitudes and is accurate [12]. Alternatively, Teardrop style small accelerometers of PCB Piezotronics could also be installed adhesively on PCBs and components. They are lightweight and exhibit minimum mass loading effects on the observed amplitude of vibration. The amplification factor Q_{Res} is computed for all the parts that experience resonance and is the basis for initiating preventive actions.

5.8.6 Initiating Preventive Actions

The computed values of Q_{Res} are tabulated for parts indicating resonant frequencies. The guidelines for initiating preventive actions for the parts experiencing vibration resonance are:

(i)   The resonant frequencies of PCB and the chassis on which the PCB is mounted should be separated by at least one octave for preventing resonant coupling [13, 14]. For example, if the resonant frequency of PCB is 150 Hz, the resonant frequency of chassis should be below 75 Hz or above 300 Hz.

(ii) The acceptable value of amplification factor (Q_{Res}) should be obtained from Steinberg equations [13, 14]. As a rule of thumb in industries, design preventive actions are needed if the computed value of Q_{Res} exceeds two [15]. Preventive actions could be initiated if the value of Q_{Res} exceeds three for resonant frequencies higher than 150 Hz as the amplitude of vibration at higher frequencies is less.

Assessment Exercises

1. Say True or False:

   (i)   Transient current stress in a circuit could induce transient voltage stress also.
   (ii) Most design-related component failures are due to transient electrical overstress and not due to operating electrical overstress.
   (iii) Electrical power overstress on components is related to thermal stress on the components.

  (iv) Improper packaging of components and modules could result in thermal overstress to components.

  (v) The supply voltages for electronic units in automobiles are derived from battery and hence the units are excluded for transient electrical overstress analysis.

  (vi) If an item is designed with resilient mounts, the item should be mounted with the resilient mounts for conducting vibration resonance overstress analysis to ensure normal mounting means.

  (vii) Vibration resonance overstress analysis detects workmanship-related defective solder joints.

(viii) Test conditions for vibration resonance overstress analysis on electronic items should be as per the Test Method Standards for conducting environmental tests.

  (ix) Electrical and thermal overstress analyses need not be performed on modules if the selection and application of components are as per the recommendations of component manufacturers.

2. Describe the method of conducting transient electrical overstress analysis on modules.
3. Describe the method of conducting vibration resonance overstress analysis on modules.
4. Explain the method of estimating worst-case electrical power overstress on discrete semiconductor devices in the modules of equipment.
5. Explain the basis for initiating preventive actions for the components that are experiencing vibration resonance stress.

Laboratory Assignments

1. Conduct overstress analyses on the bought-out products listed below. Identify the overstressed components and recommend appropriate remedial actions considering efforts and cost.

    (i)   SMPS
   (ii)   RF Power Amplifier
  (iii)   LED Lighting unit
  (iv)   Audio power amplifier
   (v)   Battery charger.

References

1. Bot, Y.: Improving product's reliability by stress de-rating and design rules check. IEEE Proc. Ann. Reliabil. Maintainability Symp. 2013. 978-1-4673-4711-2/13
2. Jang, S., Shin, M.W.: Thermal analysis of LED arrays for automotive headlamp with a novel cooling system, IEEE Trans. Device Mater. **8**, 561–564 (2008)

3. Taliyan, S.S., Sarkar, S., Biswas, B.B., Kumar, M.: Finite element based thermal analysis of sealed electronic rack and validation. In: 2nd International Conference on Reliability, Safety and Hazard, pp. 443–447. 2010
4. Application Note-AN9010, MOSFET Basics: Fairchild semiconductor corporation, Rev 1.0.5, Aprl 2013
5. Perica, G.: Ceramic input capacitors can cause overvoltage transients, application note 88, linear technology, Mar 2001
6. Application Brief on Methods for Characterizing and Tuning DC Inrush Current: Agilent technologies, USA, 5991-2778EN, July 2013
7. Foster, W.M.: Thermal verification testing of commercial printed cicrcuit boards for spaceflight. In: IEEE Proceedings Annual Reliability and Maintainability Symposium, pp. 189–195. 1992
8. Steinberg, .D.S.: Vibration analysis for electronic equipment, Wiley, New York (2000)
9. Malatkar, P., Wong, S.F.: Troy Pringle and Wei Keat Loh, Pitfalls an Engineer Needs to be Aware of during Vibration Testing, Electronic Components and Technology Conference, pp. 1887–1892. 2006
10. MIL-STD-810F: US DOD standard for environmental engineering considerations and laboratory Test, Annex B, engineering information for vibration test
11. Salvatore Liguore and David Followwell: Vibration fatigue of surface mount technology (SMT) solder joints. In: Proceedings Annual Reliability and Maintainability symposium, pp. 18–26. 1995
12. Veprik, A.M.: Vibration protection of critical components of electronic equipment in harsh environmental conditions. J. Sound Vib. **259**(1), 161–175 (2003)
13. Steinberg, D.S.: Countering the effects of vibration on board-and-chassis systems, electronics, pp. 100–102. 4 Aug 1977
14. Designing and Manufacturing Rugged COTS Assemblies: GE Intelligent Platforms, 12.10 GFT 776
15. Environmental Testing of Quartzdyne® Pressure Transducers: Design Verification Report, Quartsdyne Inc, USA, Oct 1998

Chapter 6
Reliability Evaluation of Equipment

Abstract The proto-models of electronic equipment are subjected to electrical and environmental tests as per the contractual specification of the equipment. The tests are not adequate for providing high level of confidence in the reliability of equipment. An integrated test program is designed for evaluating electrical and environmental performance and the reliability of equipment. Reliability improvement tests are superimposed with the electrical and environmental tests to the extent feasible in the integrated test program. Subjecting proto-models to the integrated test program is defined as the reliability evaluation of equipment. Performance trend analysis, random vibration, thermal shock, and accelerated burn-in are the reliability improvement tests that are explained with examples. The pre-requisites that should be satisfied for ensuring the effectiveness of reliability evaluation program are listed.

6.1 Definition of Reliability Evaluation

The proto-models of electronic equipment are subjected to electrical and environmental tests as per the contractual specification of the equipment. The tests are adequate for verifying or demonstrating compliance to the specifications, but they are not adequate for providing high level of confidence in the reliability of the equipment. An integrated test program is designed for evaluating electrical performance, environmental performance, and the reliability of equipment. Reliability improvement tests are superimposed with the electrical and environmental tests to the extent feasible in the integrated test program. Subjecting the proto-models of equipment to the integrated test program is defined as the reliability evaluation of equipment. The test program complements the overstress analyses on the proto-models. Reliability evaluation should not be misconstrued as Mean Time Between Failures (MTBF) evaluation of equipment.

© Springer International Publishing Switzerland 2015

D. Natarajan, *Reliable Design of Electronic Equipment*,

DOI 10.1007/978-3-319-09111-2_6

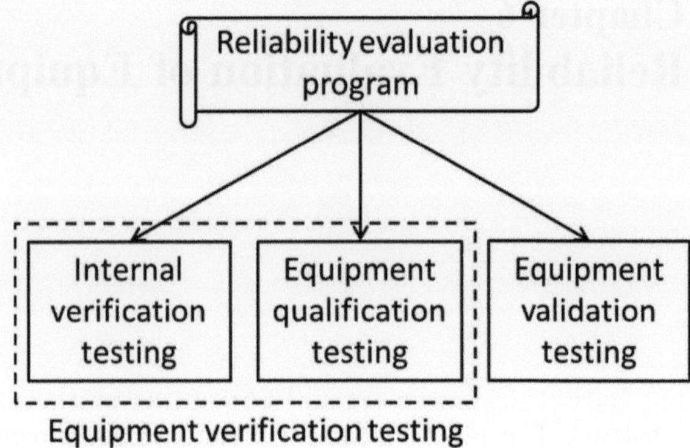

Fig. 6.1 Reliability evaluation of equipment

6.2 Reliability Evaluation Program

Reliability evaluation program for equipment is developed based on the electrical and environmental requirements of contractual documents. The reliability evaluation program of equipment is shown in Fig. 6.1 and the program contains two categories of testing which include:

(i) Verification testing

- Internal verification testing
- Equipment qualification testing

(ii) Validation testing

Internal verification testing contains value-added electrical and environmental tests for verifying contractual specifications and improving the reliability of equipment. Equipment qualification testing demonstrates compliance to the contractual specifications of equipment to customers. Validation testing of equipment is performing functional checks on equipment in actual field operating conditions. COTS items are usually mounted in relevant equipment and functional checks are performed in the equipment. Reliability evaluation program for equipment is explained with examples after mentioning the objectives and prerequisites of the program.

6.2.1 Objectives

The objectives of reliability evaluation program are common to electronic equipment and COTS items. The objectives are as follows:

(i) To force latent design and workmanship-related errors into failures
(ii) To verify compliance to contractual electrical and environmental specifications

Fig. 6.2 Prerequisites for
reliability evaluation

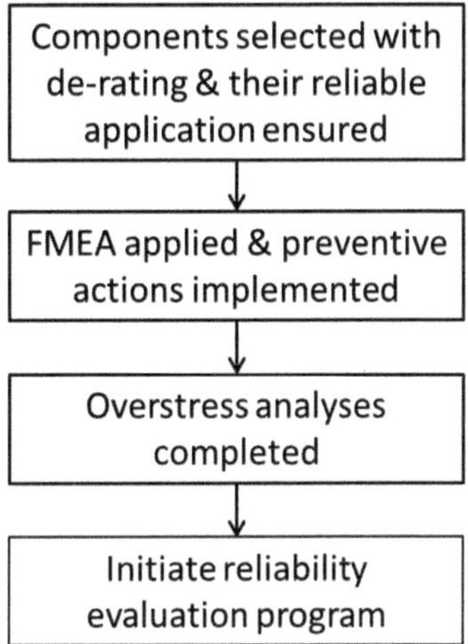

(iii) To analyze the failures, observed in the reliability evaluation program for
      identifying the root causes of the failures
(iv) To implement corrective actions to eliminate the effects of the causes
 (v) To provide high level of confidence in the reliability of equipment for
      organization and customers

6.2.2 Prerequisites

The reliability evaluation program of equipment identifies latent design error
related failures that could not be predicted during the design and development of
equipment. The program is not a means to detect obvious failures. Hence, certain
prerequisites should be satisfied for ensuring the effectiveness of reliability evaluation program of equipment. The prerequisites are as follows:

- Proper selection and application of electronic components with de-rating
- Performing FMEA and implementing the preventive actions identified in the
  analysis
- Performing electrical, thermal, and vibration overstress analysis on modules,
  units, and equipment as applicable.

Satisfying the prerequisites of reliability evaluation program enables design
teams to focus on the objectives of the evaluation program. Without the prerequisites being satisfied, too many failures would occur during evaluation, causing
delays in the program and customer dissatisfaction. The prerequisites of reliability
evaluation program are shown in Fig. 6.2.

6.3 Internal Verification Testing

Internal verification testing is the means to achieve the objectives of the reliability evaluation program and hence it is the focus of the evaluation program. The internal verification test plan of equipment contains electrical, environmental, and reliability improvement tests. The electrical and environmental tests verify compliance to contractual specifications. Reliability improvement tests are designed to force latent design errors into failures and to achieve first time acceptance of equipment in qualification testing by customers. The reliability improvement tests are superimposed on the electrical and environmental tests for minimizing time and efforts for internal verification testing. The value-added internal electrical and environmental testing is a variant of Reliability Growth Testing (RGT). More about RGT is presented in Appendix B. Internal verification test plan is explained after listing the types of latent design errors.

6.3.1 Types of Latent Design Errors

Latent design errors are those that could not be predicted during design and development processes. They creep into design processes inadvertently or due to lack of application information. Four types of latent design errors could exist in electronic designs and they are shown in Fig. 6.3. The types of errors are:

(i)   Excessive drift in electrical performance characteristics:
      Drift in electrical performance characteristics is a measure of performance reliability of equipment. Excessive parametric drift could be observed at high and low temperatures. Parametric drifts may be reversible, i.e., the performance characteristics recover to nearly initial values when the items are returned to ambient temperature or the drift might become permanent.

(ii) Components of reliability suspect circuits:
      Unknown failure modes could be associated with components used in circuits such as switching circuits that generate permissible transient electrical stress or those used in high-power RF circuits. Premature failures could occur in such components.

(iii) Defective components:
      Although the intrinsic reliability of components is quite high with present technology, chances of selecting defective components still exist.

(iv) Mechanical design errors:
      Electronic equipment could contain mechanical design errors. Examples of mechanical design errors are:

   - Improper selection of raw materials
   - Improper design of bond pads in PCBs
   - Mismatches in the thermal coefficients of materials

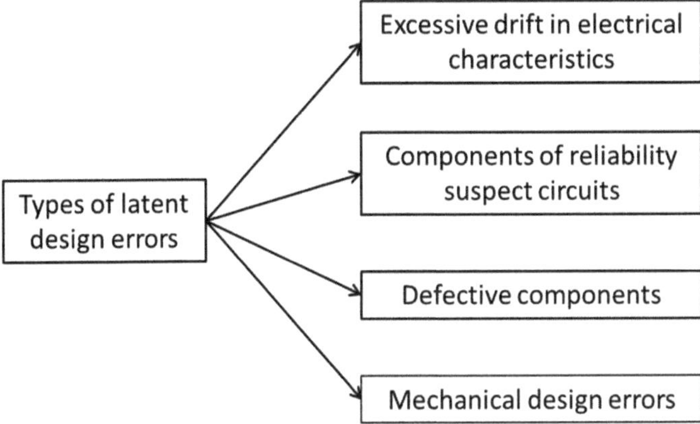

Fig. 6.3 Latent design errors

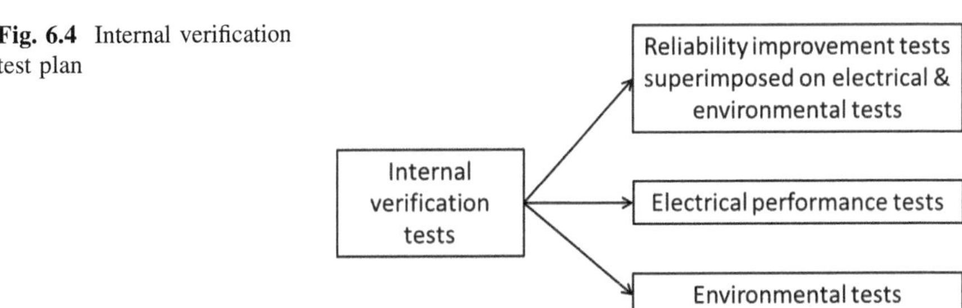

Fig. 6.4 Internal verification test plan

- Mounting of modules and units becoming loose
- Inadequacies that could not be identified in vibration resonance overstress analysis

6.3.2 Preparing Internal Verification Test Plan

The contractual document of equipment is the basis for preparing internal verification test plan. The tests in the internal verification test plan are shown in Fig. 6.4 and the tests are:

(i) Reliability improvement tests:
Reliability improvement tests are specially designed electrical and environmental tests for precipitating latent design errors into performance and component failures. The tests are superimposed with the standard environmental tests to the extent feasible. Knowledge in electronics design and environmental test engineering are essential for designing the reliability

improvement tests. Literature survey and experience are also required for finalizing realistic reliability improvement tests.

(ii) Electrical performance tests:
Electrical performance tests are specific to the equipment under reliability evaluation. The tests are conducted before, during, and after environmental tests as per the test procedures of equipment.

(iii) Environmental tests:
Environmental tests are conducted as per military or telecom or other standards, specified in contractual documents. The standards describe the methods of conducting environmental tests for various modes of transportation and operating environments. Referring or interpreting the standards and the cross-referenced documents of the standards cause delays in environmental testing. Stand-alone test procedures are prepared and approved in advance for conducting the tests to eliminate the delays.
Conducting electrical performance and environmental tests are not explained. The reliability improvement tests and the methods of superimposing the tests with environmental tests are explained with examples.

6.3.3 Reliability Improvement Tests

Electrical, thermal, and vibration stresses are applied separately or in combination to design reliability improvement tests for electronic equipment. The application of the stresses has proved to precipitate latent design errors of electronic equipment into failures. Voltage stress, power on-off, temperature, and vibration are prime environments affecting the reliability of equipment [1, 2]. Electrical stress, thermal shock with high ramp rate, vibration, and other combined environmental tests are used for improving the robustness of products and the stress levels for the tests are beyond the specified limits [3]. Multiple (cyclic) stresses and power on-off cycles are also used to search for potential failures [4]. Random vibration test exposes mechanical design errors in a short period of time using high levels of vibration [5]. On-off stressing is specified for LED lighting products [6, 7]. Reliability improvement tests that are applicable for most electronic equipment are shown in Fig. 6.5 and the tests are:

  (i) Performance trend analysis
 (ii) Thermal shock
(iii) Random vibration
(iv) Accelerated burn-in

Technical guidance for designing and conducting the tests are explained. Additional reliability improvement tests such as combined environmental tests could also be planned for the specific needs of electronic equipment.

Fig. 6.5 Reliability
improvement tests

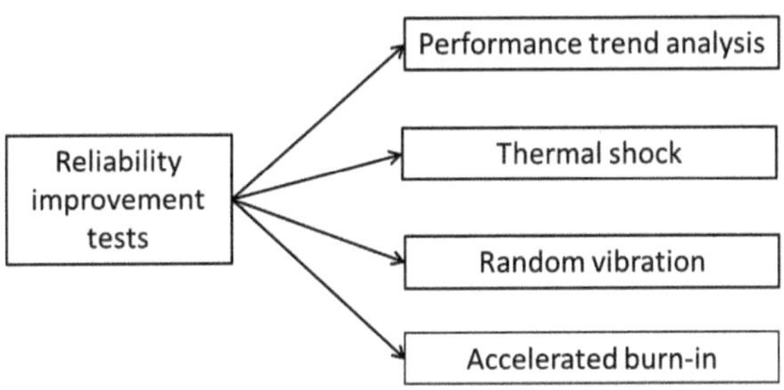

6.4 Performance Trend Analysis

Performance trend analysis concerns with the identification of measurable
parameters that affect safety and reliability of items and monitoring the parameters
at periodic intervals for initiating corrective actions [8]. The trend analysis technique could be simplified and used for improving the performance reliability of
electronic equipment. The application of the technique is integrated with internal
environmental testing.

6.4.1 Application Methodology

The electrical performance characteristics of equipment are measured at ambient
temperature before starting temperature tests. The performance characteristics are
measured during high temperature test and they are measured again after recovery
at ambient temperature. Drift, i.e., trend from initial values is computed for critical
performance characteristics measured during and after the temperature tests. The
trend in electrical performance parameters is a measure of the performance reliability of equipment. The causes of performance parameters with high values of
trend are established and corrective actions are implemented for eliminating the
causes, thereby improving the performance reliability of equipment. Performance
trend analysis is relevant for high and low temperatures and if required, the trend
of the parameters could be monitored in other environmental tests also. The
application of performance trend analysis technique at high temperature is
explained with an example.

6.4.2 Example of Application

Consider a lumped 10 MHz low pass filter with discrete inductors and capacitors.
The performance characteristics of the filter [9] at ambient temperature, at +70 °C
and after recovery are shown in Table 6.1. A trend analysis graph could also be
drawn for the characteristics.

Table 6.1 Filter performance at 70 °C

Characteristics	Limit	Initial	At +70 °C		After recovery	
			Value	Drift (%)	Value	Drift (%)
Pass band (MHz)	≥25	26	27.3	5.0	26.2	0.8
Insertion loss (dB)	≤1.5	1.3	1.32	1.5	1.28	−1.5
Return loss (dB)	≥20	26	22	−15.4	25	−3.8
Rejection (dB) (40–500) MHz	≥60	69	61	−11.6	68	−1.4

The measured values of the low pass filter at +70 °C and after recovery indicate that the filter performance is satisfactory in high temperature test. However, the trend in return loss and rejection indicate that the performance of the low pass filter is not reliable high temperature. For example, if the rejection performance of the low pass filter measures 65 dB initially, it could drop to 57.5 dB at +70 °C with the drift of −11.6 %. Hence, the performance reliability of the low pass filter needs to be improved by design as it would be difficult to control the variations during manufacturing. The root cause of the performance variations is the drift in the values of capacitors and inductors with temperature. The drift results in extending the pass band, which in turn affect return loss and rejection performance of the filter. The design corrective action for eliminating the causes of the performance trends would be to replace the capacitors with better temperature characteristics and to improve the dimensional stability of coils.

6.4.3 *Trend Analysis for In-Circuit Parameters*

Trend analysis is not limited to electrical performance characteristics. It is also applicable for the in-circuit voltages and currents of modules. Monitoring the trends of in-circuit voltages and currents contributes for the reliability improvement of equipment considerably. For example, the trends in the ripple voltage of power supply units and the DC supply current of RF Power Amplifier modules could be monitored during environmental tests. Circuit design expertise and engineering decisions are used to identify in-circuit voltages and currents for trend analysis.

6.5 Thermal Shock

Thermal shock is subjecting the item under test to cyclic temperature extremes with rapid changeover between the temperatures. Thermal shock test forces intrinsic defects in electronic components, materials, and workmanship-related solder joint defects into failures. It also exposes parts having mismatched temperature coefficients. Although the workmanship-related solder joint defects are

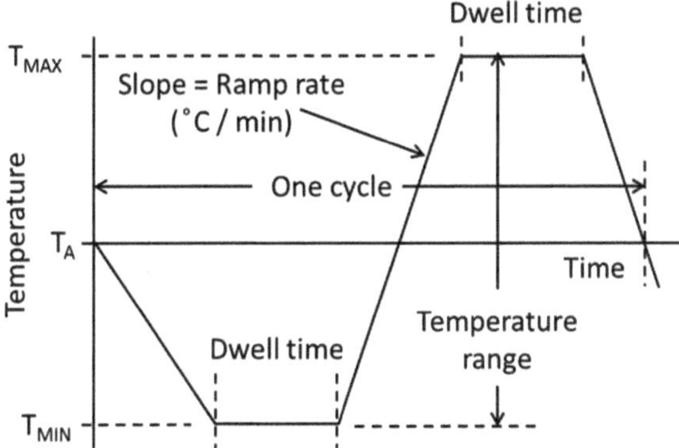

Fig. 6.6 Variables of thermal shock test

not design errors, they have to be weeded out to ensure first time acceptance in qualification testing with customers. Thermal shock test is effective in precipitating defects when the test is conducted on assembled PCBs. The test is designed to be nondestructive and hence the tested PCBs could be used for final equipment assembly. Visual examination and electrical performance tests are the means to detect defects and hence the tests are conducted on PCBs before and after thermal shock. The test results are used to improve the effectiveness of thermal shock test by modifying test conditions.

6.5.1 Test Conditions

Test conditions for thermal shock are decided by the variables associated with the test. The variables as shown in Fig. 6.6 are:

 (i) Temperatures range, i.e., the difference between upper and lower temperatures
 (ii) Rate of temperature change or the ramp rate (°C/min)
(iii) Dwell time at lower and upper temperatures
(iv) Number of thermal shocks.

Test conditions for thermal shock test should be decided to ensure that the test is effective for precipitating the defects of PCBs; otherwise thermal shock test would become benign. For example, thermal shock test with the upper temperature is set at +55 °C and the lower temperature is set at −10 °C is not capable of precipitating the defects of PCBs as the temperature range is 65 °C, which is inadequate. In another case, assume that thermal shock test for PCBs is designed with the upper temperature set at +100 °C and the lower temperature set at −40 °C. The temperature range of the shock test is 140 °C, which is capable of precipitating the defects of the PCBs. When the PCBs are subjected to thermal shock test between −40 and +100 °C, assume that the components, materials, and solder joints of the PCBs experience temperature variation between $-14\ ^\circ$C and $+63\ ^\circ$C due to thermal lag and other factors. As the parts of the PCBs experience thermal shock with temperature range of 77 $^\circ$C only, the thermal shock test is benign, i.e., not capable of precipitating the defects of the PCBs.

6.5.2 Temperature Range

Temperature range should be high for precipitating the defects of PCBs under test. The storage temperature limits of electronic equipment could be used as temperature limits for thermal shock test. The typical storage temperature limits are -55 and $+125\ ^\circ$C (temperature range, 180 $^\circ$C) for military products and -40 and $+85\ ^\circ$C (temperature range, 125 $^\circ$C) for consumer products [10]. The temperature ranges are acceptable for thermal shock test. In general, the temperature limits for thermal shock test should not exceed the maximum storage temperature limits of electronic components, specified by manufacturers.

6.5.3 Ramp Rate

Ramp rate is the rapidity of changeover between temperature extremes and it is expressed in degree centigrade per minute. Ramp rate decides whether the item under test is experiencing thermal shock or thermal cycling. Ramp rate for thermal shock is usually greater than 20 $^\circ$C and it is less than 20 $^\circ$C for thermal cycling [10]. High ramp rates are effective in exaggerating stresses associated with mismatches in coefficients of thermal expansion in metals, plastics, and adhesives [11]. Tests with excessively high ramp rates actually could be ineffective in precipitating flawed solder joints into failures [12]. Ramp rate in excess of 30 $^\circ$C is required for reliability testing of surface mount solder attachments [13].

Simulation of excessive ramp rates is expensive. Most of the electronic components are tested for thermal shock in the temperature range, -65 to $+125\ ^\circ$C with a ramp rate of 40 $^\circ$C approximately as per military standards [14]. Ramp rates between 30 and 40 $^\circ$C could be considered for conducting thermal shock test on PCBs and modules.

6.5.4 Dwell Time and Number of Shocks

High dwell time is effective in exposing solder joint defects. Creep is identified as the key mechanism in causing solder failures and high dwell time is required for creep to occur [12]. A dwell time of 5 min is used at high and low temperatures as per current practice in industry [12]. Industries use a dwell time of 5 to 15 min and higher dwell time could be considered for PCBs populated with heavier components such as ferrite pot cores to allow temperature stabilization.

Most of the electronic components are subjected to five or twenty-five thermal shocks in qualification testing as per military standards. As the thermal shock test is designed to be nondestructive, five thermal shocks could be considered adequate for PCBs and modules.

6.5.5 Loading PCBs in Test Chamber

Thermal shock test chamber with single or double compartments is used for conducting thermal shock test and PCBs are loaded into the chamber. The test conditions of thermal shock test are set in the chamber and they represent the conditions of air circulating inside the chamber. Thermal shock test is effective in exposing the defects of PCBs only if the parts (components and solder joints) of the PCBs experience the same temperature conditions of air circulating inside the chamber.

Assuming thermal shock test chamber is capable of achieving the desired ramp rate, the method of loading of PCBs in chamber compartment decides the effectiveness of temperature coupling between the parts of PCBs and circulating air. The PCBs should be mounted vertically in a test fixture with appropriate spacing for air circulation. The manufacturers of test chambers specify the size and maximum number of PCBs that could be loaded in chamber compartments. Their recommendations should be adhered for achieving effective temperature coupling between the parts of PCBs and circulating air. If required, thermocouples could be attached to PCBs to confirm that the temperatures are within tolerances as per IEC 60068-3-5:2001, Supporting documentation and guidance-Confirmation of the performance temperature chambers, and the set ramp rate.

6.6 Random Vibration

The vibration experienced by equipment during transportation or operation is generally random in nature. It is realistic to test equipment with random vibration instead of sinusoidal vibration. Random vibration test exposes mechanical design inadequacies and errors (Sect. 6.3.1) in the design of electronic equipment. Random vibration test could be conducted on finished COTS items, but it is recommended to conduct the test on the functional units of equipment. The test conditions for random vibration test on functional units are different from those for qualification testing on finished equipment and they are designed to be nondestructive. The characterization of random vibration should be understood for deciding the test conditions.

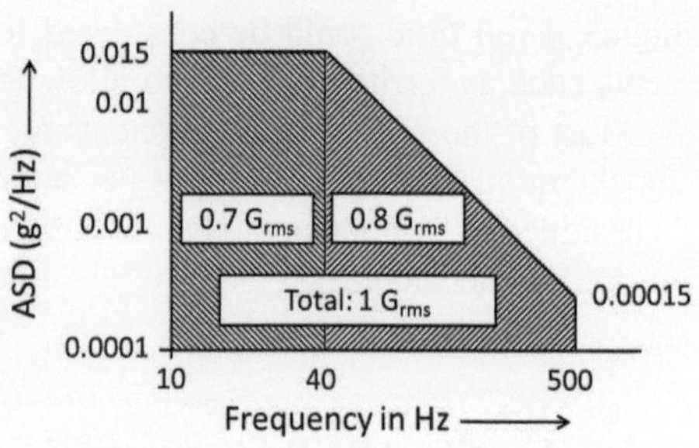

Fig. 6.7 Example of random vibration profile

6.6.1 Characterization of Random Vibration

Random vibration is characterized by frequency range and vibration level. Minimum and maximum frequency limits in Hz are specified for frequency range. Acceleration Spectral Density (ASD) in g^2/Hz specifies the vibration level for the frequency range. Different vibration levels are specified within the frequency range of most vibration profiles. The frequencies at which the vibration levels change are crossover frequencies. An example of random vibration profile from MIL-STD-810G is shown in Fig. 6.7 indicating frequency range and vibration levels at crossover frequencies.

The random vibration profile, shown in Fig. 6.7, has two frequency bands within the frequency range and one crossover frequency. For each frequency band, the area contained by the (Frequency—ASD) curve could be calculated and the square root of the area is expressed in gravity unit, G_{rms}. The gravity unit of a frequency band indicates the intensity or the energy level of random vibration in the frequency band. The total energy level of random vibration for the entire frequency range could also be calculated. The G_{rms} levels for the frequency bands and the total G_{rms} level for the entire frequency band. The method of computing G_{rms} level from ASD levels for random vibration profiles is explained in Appendix C.

6.6.2 Test Conditions

NASA-STD-7001A, Payload Vibroacoustic Test Criteria, specifies overall 6.8 G_{rms} in the frequency band (20–2000) Hz for one minute for acceptance testing of electronic hardware weighing less than 50 kg. MIL-STD-810G, Environmental Engineering Considerations and Laboratory Tests, specifies 7.7 G_{rms} the frequency band (20–2000) Hz as minimum integrity exposure for electronic hardware weighing less than 36 kg. The military standard specifies 500 Hz as the maximum frequency for the transportation of items by trucks. Hence, the test condition for conducting random vibration on electronic equipment and COTS items is decided as (20–500) Hz with 4.4 G_{rms}. The vibration profile is shown in Fig. 6.8. As the

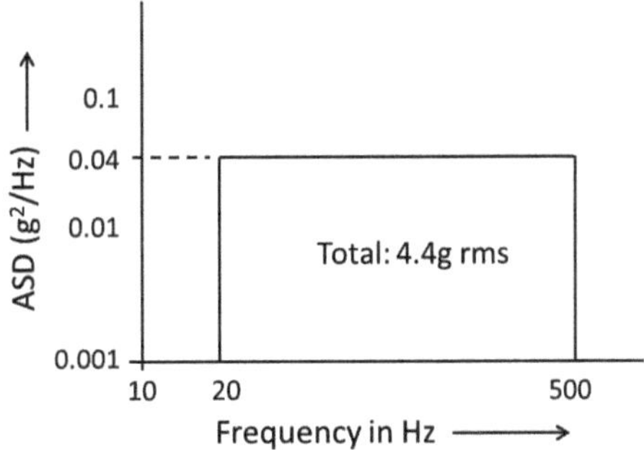

Fig. 6.8 Random vibration test profile

vibration test needs to be nondestructive, the duration of test could be set between 5 and 10 min in each plane based on vibration trials on the item under test.

6.6.3 Conducting Random Vibration Test

Visual examination is the prime means of detecting exposed defects after vibration and the detection is supported by electrical performance checks. Hence, both the tests are conducted on the item under test before and after random vibration. A resonance-free fixture is mounted on the table of vibration machine. The unit under test is hard mounted without using resilient mounts on the vibration fixture for subjecting the unit for random vibration. The test is conducted in each of the three perpendicular planes.

6.7 Accelerated Burn-in Test

Accelerated burn-in test is operating electronic equipment at stress levels higher than its normal stress levels specified in the contractual documents of the equipment for a short period of time. In general, accelerated testing technique provides a shortcut method to investigate the reliability of electronic items and it could be conducted with or without the knowledge of failure mechanisms of the items [15]. The test is effective if designed with the knowledge of failure mechanisms of electronic components. For example, higher DC current and junction temperature accelerates the failure of RF power transistors [16] and this information is useful to design an accelerated test for RF Power Amplifiers.

Accelerated burn-in test forces marginal designs into electrical performance and component failures, providing opportunities for improving the reliability of electronic equipment. The test could be integrated with high temperature test with extended exposure time during internal environmental testing. Supply voltage, input signals, and output load conditions are set to ensure maximum operating electrical stress for the item under accelerated burn-in test. Usage on-off stressing is also simulated during the test. Identifying usage on-off stressing circuits and deciding test temperature and duration are explained after mentioning the basic requirements of accelerated burn-in test. The basic requirements are common to all electronic equipment and COTS items.

6.7.1 Basic Requirements

Electrical and thermal stresses are applied on the item for conducting accelerated burn-in test. The stresses degrade the performance of components during the test. The levels of electrical and thermal stresses for accelerated burn-in test should be designed to ensure that the rate of degradation process in electronic components remains linear with the stresses. Linearity in the degradation process is ensured by not exceeding maximum ratings specified by manufacturers for components during accelerated burn-in test. If stress levels higher than maximum ratings are applied, the rate of degradation process becomes nonlinear, i.e., unpredictable with the stresses. Performance outside the specifications of data sheet is not guaranteed and it results in unexpected failures and new failure mechanisms [17]. Such failures are bound to misguide the reliability improvement process of equipment.

6.7.2 Usage On-Off Stressing

Controls are available in the front panel of electronic equipment for users to select and operate the functions of the equipment. The operation of the controls enables or disables appropriate electronic circuits in equipment. Enabling or disabling electronic circuits is equivalent to switching on or off the circuits. Selecting transmit or receive function in communication equipment using a microswitch or PIN diode is an example usage on-off stressing. On-off stressing is specified for LED lighting products [6, 7]. On-off operating life test is conducted for the qualification testing of high power semiconductors with specific drive and snubber circuits [18]. Examples of the circuits that need to be considered for on-off stressing in electronic equipment are:

(i) Power switching circuits
(ii) Reliability suspect switching circuits that generate permissible transient voltages and currents (Sect. 5.5.2)

Performance trend analysis (Cl. 6.3.4) is integrated with on-off stressing test by measuring performance electrical characteristics of switching modules. The final measured values are compared with initial values for implementing reliability improvement actions.

Table 6.2 Deciding test temperature for accelerated burn-in test

Transistor/module	Worst-case operating junction temperature of the transistors (°C)		$T_{j\text{-max}}$ (°C)
	Equipment at 50 °C	Equipment at 75 °C	
TR$_2$/Mod-1	130	155	175
TR$_4$/Mod-2	140	165	200
TR$_1$/Mod-3	175	200	200
TR$_4$/Mod-4	110	135	175
TR$_1$/Mod-5	135	160	175

On-off stressing is mostly relevant for semiconductor devices. The test conditions of intermittent operation life test as per MIL-STD-750F, Test Methods for Semiconductor Devices, are specified in data sheets. The same could be applied for conducting usage on-off stressing test on switching circuits. The duty cycle for usage on-off stressing test is fixed as five minutes power-on and five minutes power-off for 500 cycles.

6.7.3 Test Temperature and Duration

Test temperature for accelerated burn-in test is primarily based on the worst-case operating junction temperature of semiconductor devices. They are obtained from thermal interaction stress analysis (Cl. 5.7.2) on equipment. The method of deciding test temperature for accelerated burn-in test on equipment is explained with an example.

Assume hypothetical equipment with five modules and the maximum operating ambient temperature of the equipment is 50 °C. The worst-case operating junction temperatures of the transistors of the five modules are shown in Table 6.2, assuming that the equipment is operating 50 °C. The absolute maximum junction temperatures of the semiconductor devices ($T_{j\text{-max}}$) are also shown in the Table.

If the operating ambient temperature of the hypothetical equipment is increased from 50 to 75 °C, the expected worst-case device operating junction temperatures would also increase proportionately, ignoring the computational errors. The increased temperatures are also shown in the Table 6.2. The transistor, TR$_1$ of Mod-3, is operating at its maximum junction temperature and other transistors are operating below their maximum temperature. Theoretically, burn-in test could be conducted at the accelerated temperature of 75 °C for the equipment. But it would be safer to conduct the accelerated burn-in test on the equipment at 65 °C to prevent thermal runaway of semiconductor devices. The duration of accelerated life test could be set to that of standard burn-in test and it is 48 or 96 h.

6.8 Cause Analysis and Corrective Actions

Both functional and component failures might occur during internal verification testing program. Failures might also be observed during equipment qualification and validation testing. All failures should be recorded. The observed failures are analyzed to understand the root causes of the failures. Corrective actions are implemented for preferably eliminating the root causes. All failures are caused and they do not happen. Hence, root cause analysis and implementing corrective actions are critical to ensure success in the reliability evaluation program of equipment. The effectiveness of the corrective actions should also be monitored and recorded. Failure recording, cause analysis, and implementing corrective actions are explained in Chap. 7, Failure Data Management.

Assessment Exercises

1. Say True or False:

   (i)   Reliability improvement tests are part of internal verification testing and hence they are not conducted in the presence of customer or their representatives.
   (ii)   The design actions to eliminate the causes of failures observed in internal verification testing are preventive actions.
   (iii) Electrical and thermal overstress analyses on the proto-models of equipment prevent component failures with certainty.
   (iv)   Drift in electrical performance characteristics at high temperature is a measure of the reliability of electrical performance.
   (v)   Vibration resonance overstress analysis and random vibration test precipitates similar mechanical defects.
   (vi)   Number of temperature cycles decides whether an item is experiencing thermal shock or thermal cycling.
   (vii) Random vibration test is applicable for reliability improvement and qualification testing of electronic items.
   (viii) Random vibration profile having three vibration levels in the frequency band of the profile has three crossover frequencies.
   (ix)   The test conditions of random vibration test for internal reliability improvement should not exceed the test conditions for qualification testing of equipment.

2. What are the prerequisites that should be satisfied before initiating reliability evaluation program on equipment? Why they should be satisfied?
3. List the types of latent design errors and explain them with examples.
4. Explain two reliability improvement tests with examples.

5. Explain the test conditions that could result in benign thermal shocks to the components of PCBs.
6. Explain the variables associated with thermal shock test.
7. Why loading PCBs in thermal shock test chamber is important?
8. Explain how random vibration is characterized.
9. Explain the basic requirements of accelerated burn-in test on equipment.

Research Assignments

1. Design a reliability improvement test program for the electronic modules listed below. The program should consist of Performance trend analysis including in-circuit parameters, Thermal shock, Random vibration, and Accelerated burn-in tests. Additional product specific reliability improvement tests could also be included. The test conditions should be supported by documentary evidences. The required inputs for designing accelerated burn-in test could be assumed.

    (i)   SMPS
    (ii) RF power amplifier
    (iii) LED lighting unit
    (iv) Audio power amplifier
    (v)   Battery charger.

Laboratory Assignments

1. After designing the reliability improvement test program on the products listed below, conduct the tests on the bought out products. Tabulate the failure data observed in the tests.

    (i)   SMPS
    (ii) RF power amplifier
    (iii) LED lighting unit
    (iv) Audio power amplifier
    (v)   Battery charger.

References

1. Dishman, J.C.: COST saves dollars critical operating stress testing methods for early design analysis. IEEE Proceedings of Annual Reliability and Maintainability Symposium, pp. 186–190 (1989)
2. Caruso, H.: An overview of environmental reliability testing. IEEE Proceedings of Annual Reliability and Maintainability Symposium, pp. 102–109 (1996)

3. Zunzanyika, X.K., Yang, J.J.: Simultaneous development and qualification in the fast changing 3.5" hard-disk-drive technology. IEEE Proceedings of Annual Reliability and Maintainability Symposium, pp. 27–32 (1995)
4. Perera, U.D.: Reliability of mobile phones. IEEE Proceedings of Annual Reliability and Maintainability Symposium, pp. 33–38 (1995)
5. Conner, S.A., Watts, P.E.: Transformation of thermal ink-jet product reliability strategy. IEEE Proceedings of Annual Reliability and Maintainability Symposium (2012) 978-1-4577-1851-9/12
6. Specification of LED Lighting Products, Guidelines Produced Under the Umbrella of the Lighting Industry Liaison Group and Endorsed by its Member Organisations, June 2011
7. LED Luminaire Lifetime Recommendations for Testing and Reporting, Solid-State Lighting Product Initiative from the US Department of Energy and Next Generation Lighting Industry Alliance, Second Edition, June 2011
8. Crawford, J.L., Wienstock, R.: The NASA trend analysis program. IEEE Proceedings of Annual Reliability and Maintainability Symposium, pp. 25–30 (1990)
9. Natarajan, D.: A Practical Design of Lumped Semi-lumped and Microwave Cavity Filters. Springer, Berlin (2013)
10. IPC 9701:2002, Performance Test Methods and Qualification Requirements for surface Mount Solder Attachments
11. Caruso, H.: The ESS muddle: physics versus relics. IEEE Proceedings of Annual Reliability and Maintainability Symposium, pp. 233–241 (1995)
12. Kallis, J.M., Hammond, W.K., Curry, E.B.: Stress screening of electronic modules: investigation of effects of temperature rate of change. IEEE Proceedings of Annual Reliability and Maintainability Symposium, pp. 59–66 (1990)
13. IPC-SM-785:1992, Guidelines for Accelerated Reliability Testing of Surface Mount Solder Attachments
14. MIL-STD-202G:2002, Test Method Standard Electronic and Electrical Component Parts
15. Hu, J.M., Barker, D., Dasgupta, A., Arora, A.: Role of failure-mechanisms identification in accelerated testing. IEEE Proceedings of Annual Reliability and Maintainability Symposium, pp. 181–188 (1992)
16. Martinez, E.C., Miller, J.: RF power transistor reliability. IEEE Proceedings of Annual Reliability and Maintainability Symposium, pp. 83–87 (1994)
17. Watson, J., Castro, G.: High-temperature electronics pose design and reliability challenges, Analog Dialogue, vol. 46. Analog Devices, USA (2012)
18. Reinhard, S.: Reliability and Testing, Section-4, ABB Semiconductors AG

Chapter 7
Failure Data Management

Abstract In spite of applying reliability techniques during the design and development of equipment, failure data could still arise in the life cycle of electronic equipment. The failures are categorized as electrical performance, component and mechanical failures. Organizing a failure data management system (FDMS) to handle the failure data is presented. Practical considerations for the root cause analysis of component failures and classifying the origin of the causes as design error, manufacturing process induced failure and defective component are explained.

7.1 Failure Data

The engineering techniques (selection and applications of components, de-rating, and FMEA) are applied to design reliable electronic equipment. The reliability of design is evaluated by conducting overstress analyses, verification, and validation tests on the proto-models of equipment. The application of engineering techniques and the reliability evaluation tests on the proto-models contribute considerably for the reliability of equipment. However, no test sequence or pattern of testing can guarantee detection of all deficiencies [1]. Hence, failure data arises during the life cycle of electronic equipment. Failures are opportunities for learning, and a system is needed to collect, organize and analyze failure data.

7.1.1 Categories of Failure Data

Failure data of electronic equipment are classified into three categories as shown in Fig. 7.1. The classification is beneficial for managing failure data for cause analysis. The categories are:

© Springer International Publishing Switzerland 2015

D. Natarajan, *Reliable Design of Electronic Equipment*,

DOI 10.1007/978-3-319-09111-2_7

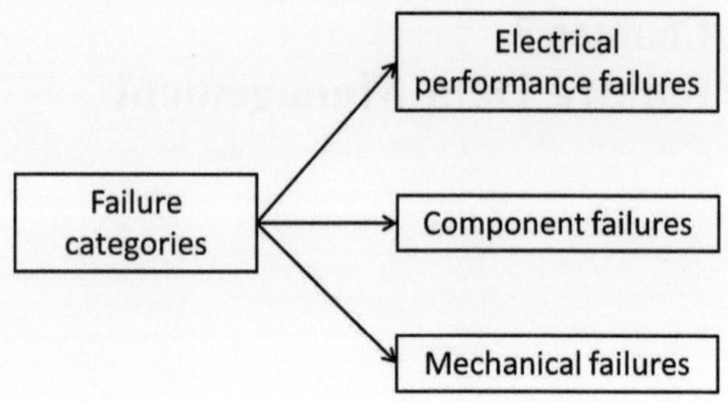

Fig. 7.1 Categories of failures

  (i) Electrical performance failures:
Poor signal-to-noise ratio and high distortion in carrier frequency are examples of electrical performance failures.

 (ii) Mechanical failures:
Loosening of cable-connector interconnections and ICs dislodging from sockets are examples of mechanical failures.

(iii) Component failures:
Base-emitter junction of transistor short and capacitor open are examples of component failures.

7.1.2 Sources of Failure Data

Failure data could arise during the early stages of development of equipment such as performance testing of first proto-models of PCBs. Such failures are analyzed and corrective actions are implemented by circuit designers. A formal system is required for handling failure data arising from the sources listed below:

  (i) Internal verification testing
 (ii) Qualification and validation testing
(iii) Manufacturing of equipment
(iv) Usage of equipment

The formal system is referred as Failure Data Management System (FDMS). The acronym, FDMS, is used throughout the chapter to keep the presentation simple.

7.2 Failure Data Management System

Figure 7.2 shows FDMS with inputs and outputs. Failure data reports are the inputs to FDMS. Corrective and preventive actions are the outputs of the system. Corrective actions are used to improve the quality of product design or the quality

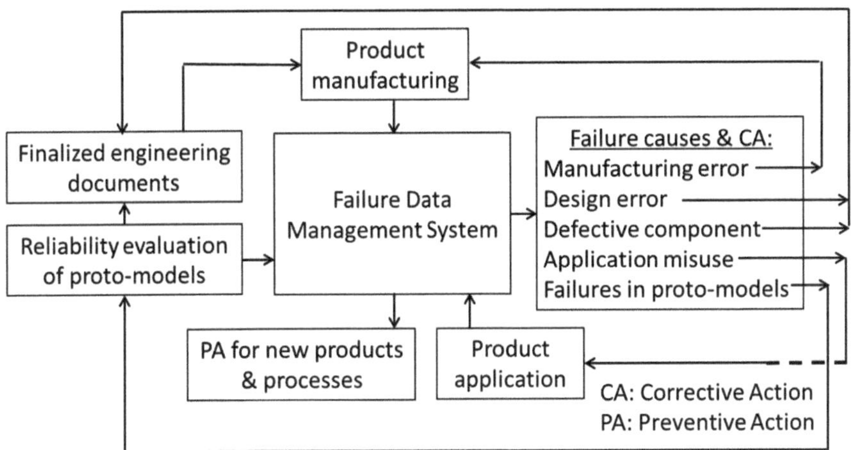

Fig. 7.2 Inputs and outputs of FDMS

of manufacturing process design or both thus providing a closed loop management system. Preventive actions are information feed-forward for new designs. The activities to derive the outputs of FDMS from inputs could be broadly classified as:

- Administrative requirements
- Analysis of failures

7.2.1 Administrative Requirements

Failure analysis for identifying the root causes of failures is time-consuming. Efforts for the analysis could be minimized by controlling failure data from their receipt to formally closing the analysis. The administrative controls on failure data are designed considering the requirements of products, organization, and customer base. The elements of administrative controls are listed below with brief explanation:

(i) Receipt of failure data:
    Failure information could be received in the form of a note or in a format. If it is received verbally from customers, the same should be documented.
(ii) Verifying failure data:
    The correctness of failure description, equipment reference, part number, part description, circuit reference, etc. should be verified.
(iii) Preliminary analysis:
    Preliminary analysis should be conducted on failure information to conclude whether the information is a false alarm or not. Originators could be communicated appropriately to minimize the loading of FDMS.

(iv) Categorizing failure data:
   Failure data from internal sources and customers should be handled separately. Provisions are also required to handle the categories of failure data (Sect. 7.1.1) and to handle the data from various sources (Sect. 7.1.2)
 (v) Trend analysis reports:
   Failure trends are useful for identifying the root causes of failures faster.
(vi) Status reports:
   Status reports are the controls of FDMS. They are used to set priorities for analyzing failure data, to monitor the progress, and to organize additional resources for increasing the pace of failure analysis.
(vii) Closing failure analysis:
   After identifying the root causes of failure, corrective actions are identified. Provisions are needed to monitor the implementation of corrective actions and the effectiveness of the actions before formally closing failure analysis.

FDMS should be computerized for satisfying administrative requirements and using the outputs of the system to improve products and processes.

7.2.2 Analysis of Failures

Failure data is analyzed for identifying the root causes of failures and recommending corrective actions to eliminate the effects of the causes. Failure analysis is the most important activity and it decides the success or failure of FDMS. The three categories of failures (Sect. 7.1.1) are electrical performance, mechanical, and component failures. Analysis of electrical performance failures is not explained as the failures are specific to equipment. The solutions to mechanical-related failures are obvious, and they are also not explained. The methods of analyzing electronic component failures are explained.

Controlling FDMS and the analysis of failure data are usually entrusted to internal QA team to eliminate bias in the analysis. It is desirable to achieve 100 % success in failure analysis. In practice, there are practical difficulties such as nonavailability of failed components and adequate information for confirming the existence of failures. The cases, where failure analysis could not be completed, are identified for continuing the analysis with the receipt of additional information.

7.3 Cause Analysis on Component Failures

Component failures could be observed during reliability evaluation, manufacturing, and usage period. The failures are recorded and analyzed for:

 (i) Identifying root causes of the failures
 (ii) Categorizing the origin of the root causes as:

- failures induced by design errors
- failures induced by manufacturing processes
- failures induced by defective components

(iii) Recommending multiple corrective actions to eliminate the effects of the causes

Root causes are the fundamental causes of failure and corrective actions would be effective only if the root causes are addressed. Multiple corrective actions are suggested by QA so that the most practical corrective actions are implemented by design and manufacturing teams. Conducting failure analysis is explained for the three categories of component failures. The general procedure for analyzing component failures for root causes is presented in the form of flow chart.

7.3.1 Flow Chart for Failure Analysis

The steps in identifying root causes of failures, categorizing the origin of the root causes and recommending multiple corrective actions are shown in the form of flow chart in Fig. 7.3. The steps are explained in detail, and examples are provided where required.

Fig. 7.3 Failure analysis flow chart

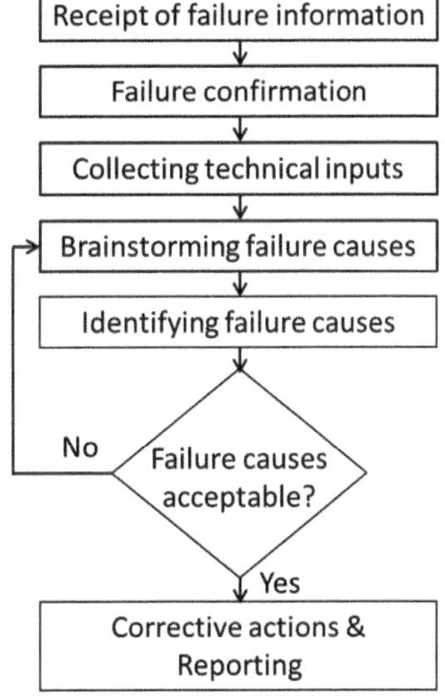

7.3.2 Failure Confirmation

Certain practical aspects should be considered by design and manufacturing departments before reporting failures to QA. Components might fail due to inadvertent errors of operators such as application of excess voltage and erroneous inputs to electronic modules. Failure data should be scrutinized by competent engineers before reporting to QA. When requests for failure analysis are received, on-site preliminary failure analysis is also conducted by QA to confirm the existence of component failures before formally recording the failures in FDMS for avoiding infructuous efforts. The preliminary failure analysis involves:

   (i)   Collection of all failed components
  (ii)   Informal discussion with technicians and engineers
 (iii)   Examining failure trends
  (iv)   External visual examination and electrical testing of failed components
         (refer Sect. 7.6.2)
   (v)   Correlating the discussions, examinations, and testing to confirm the
         reported failure

7.3.3 Technical Inputs

Stress exceeding strength causes the failure of electronic components and it is the basic technical input for finding the root causes of failures. Additional technical inputs are necessary to find the root causes of component failures and they are:

   (i)   Understanding the functioning of circuits:
         The schematic diagrams of electronic modules in which the failed com-
         ponents are used and the application notes of semiconductor devices are
         collected to understand the functioning of circuits. Discussion with
         designers complements the understanding of the circuits.
  (ii)   Understanding components:
         A brief knowledge on the design and construction of components is
         required. The information is available in the catalogues of leading inter-
         national manufacturers of electronic components.
 (iii)   Guidance for failure analysis:
         Handbooks and application notes of component manufacturers provide
         valuable information for failure analysis [2].
  (iv)   Understanding environmental effects on components:

         •   Temperature and thermal shock are some of the environmental stresses
             that could induce failures. The relationship between environmental
             stresses and their effects on components should be understood.

- Military standards [3–5] on the test methods of components describe the purpose of tests. The clauses describing the purpose of environmental tests in the military standards and journals on physics of failures are sources for understanding environmental effects on electronic components.
- Guidance for the mounting and soldering of components with manufacturing process failure modes and causes are available in the catalogues of component manufacturers. Information on the environmental effects could be derived from the catalogues.

7.3.4 Brainstorming for Failure Causes

Failure analysis is a technically challenging task. It requires collective discussions and efforts to identify failure causes. The collective discussion is termed as brainstorming. Information gathered through informal discussions with the engineers and technicians of design and manufacturing departments is discussed in brainstorming sessions. The approach for identifying failure causes is discussed and documented. The inputs for finalizing the approach are:

  (i)   Examining previous reports on similar failures
 (ii)   Listing examination and tests on failed components
(iii)   Overstress stress analyses on the modules in which the failed components are used
 (iv)   Studying manufacturing processes
  (v)   Microscopic internal examination of failed components
 (vi)   Listing possible causes of failure
(vii)   Listing tests on new components

7.3.5 Practical Considerations

Failures of components might be caused by design errors or by manufacturing processes or by defects in the components. However, components fail mostly due to design errors or manufacturing process induced stresses as the intrinsic reliability of electronic components is high with present technological advancements. Although it's a common practice to apply de-rating to components during design, the portion of failures due to overstresses is still very significant, approximately 25 % [6]. Failure analysis always begins assuming that design error has caused the failure of component, then studying manufacturing processes and finally evaluating components for defects. The suggested order could be modified if there is conclusive evidence of defects in components.

A report is prepared by QA after completing analysis indicating the root causes of failure and corrective actions. The failure report is reviewed by design and manufacturing departments. In order to ensure the acceptance of the report after the review, the failure causes should be established by two or more conclusive evidences, derived from different types of examination and tests. For example, assume that the failure of many new components is observed in a simulated switching life test for identifying root causes. The observation is one evidence only although many components had failed in the life test. Additional evidences should be generated through different type of failure analysis tests such as microscopic internal examination and overstress analysis. Generating two or more evidences should be planned in brainstorming sessions for confirming the causes of component failures.

Suggestions for corrective actions might emerge in brainstorming session. They should be discouraged at this stage as the root causes of failure are not understood. Failures might recur if corrective actions are implemented without understanding the root causes of failures. In the interest of identifying failure causes faster, the details of the failure analysis approach should be kept confidential by QA until the root causes are established.

7.4 Analysis of Failures Induced by Design Errors

Identifying the causes of component failures induced by design error is explained with an example. An electronic module, EM_1, is assumed to have failed during manufacturing. Troubleshooting operations on the module zeroed on the MOSFET power transistor, CP_1. It is further assumed that five MOSFET Power transistors had failed, confirming the existence of component failure. Discussion with the technicians of manufacturing department revealed that failure of the MOSFET was not observed during the testing of the module, EM_1, and the failure was observed only when testing the unit, EU_1. The module, EM_1, is part of the unit, EU_1. The failed MOSFETs, new MOSFETs, one new module, EM_1 and one new unit, EU_1, were collected.

Electrical tests on the failed MOSFET devices showed open-circuit in the devices. Operating electrical and thermal overstress analyses were conducted on the module, EM_1, and on the unit, EU_1. All components of the module and the unit were operating at stress levels below their maximum ratings with acceptable de-rating.

Transient electrical overstress analysis was conducted on the module, EM_1, and no transient current was observed in the module. The overstress analysis was conducted on the unit, EU_1, and excessive transient current through the MOSFET, CP_1, was observed. When the switching operations on the unit, EU_1, were repeated at the rated high temperature of the unit, the MOSFET failed. The failed devices were de-capped for microscopic (40–100X) internal examination. The bond wires had fused-out. When excessive current flows through a bonding wire, the wire may heat up and fuse [7]. The three conclusive evidences confirm that the failure of the MOSFET, CP_1, is caused by transient electrical overstress, which is a design error.

7.5 Analysis of Failures Induced by Manufacturing Processes

Handling, soldering, and post-cleaning operations after soldering are the prime manufacturing processes that apply environmental stresses on components, inducing failures. The manufacturing processes are examined in micro-steps and they are documented in suitable format for understanding the causes of failures. After understanding the causes, the manufacturing processes are applied or simulated on new components to conclude on the causes of the component failure. Technical information is provided for identifying the root causes of component failures induced by manufacturing processes.

7.5.1 Sources of Overstress

Improper handling of CMOS devices could cause Electro-static discharge (ESD)-related failures. References, [2, 7] contain additional information regarding ESD-related failures. Characteristic impedance of coaxial cables might be altered by not observing minimum bending radius requirements during handling. Maximum limit of torque specified by component manufacturers could be exceeded causing the failure of end stops in rotary wafer switches and coupling nuts in coaxial connectors.

Components experience high temperature in the order of 250 °C for few seconds in all types of soldering processes. Errors in the controls of soldering processes could cause the failure of components. The plastic package of ICs could crack during soldering due to thermal stress [7]. De-tuning of surface mounted RF Filters and permanent drift in the characteristics of components could be observed. De-lamination of PCBs and failure of bond pads in PCBs are also induced by soldering process. Excessive number of soldering and de-soldering cycles could degrade the reliability of components. Post-cleaning operations after soldering generally use volatile solvents such as isopropyl alcohol or trichloroethane for removing flux residues in assembled PCBs. If cleaning operation is performed on PCBs immediately without allowing recovery to ambient temperature, miniature components like chip ceramic capacitors fail due to dielectric cracking as they experience liquid thermal shock [8].

7.6 Confirming Defective Components

Military standards are well established and exist practically for all categories of electronic components. Microcircuits, discrete Semiconductor devices, RF Coaxial connectors and RF Coaxial cables are a few examples of the categories of components. The standards could be used for confirming failures caused by defective components. The findings on defects are repeatable and are acceptable to component manufacturers. A brief introduction to the military standards for electronic component categories is provided before deriving the test program.

7.6.1 Military Standards for Electronic Components

Military standards are organized as general specifications and specification sheets (slash sheets). The electrical and environmental performance requirements of each category of components are specified in general specification applicable for the category of component. Large types of parts exist in each category of components considering function, characteristics, rating, and mounting provisions. Slash sheets specify the unique requirements of parts and they are attached to the general specifications. For micro-circuits, the nomenclature, Standard Micro-circuit Drawing (SMD) is used instead of slash sheet. There are also separate set of Standards for Test Methods [3–5], describing the purpose and method of conducting the tests, listed in general specifications.

Examination and tests are conducted on failed parts for deriving a test program from military standards to confirm defective components. The data of failed components are correlated with the tests in general specification, slash sheet, and the purpose of the tests described in the Standard for Test Methods for deriving the test program. If military slash sheet does not exist for a failed part, the slash sheet should be prepared using data from part manufacturer and the nearest slash sheet of military standard for the failed part. Deriving test program is explained for ICs, semiconductor devices, and passive components.

7.6.2 Examination and Tests on Failed Components

Samples of failed parts are necessary to formulate a test program for confirming the defects of the parts. Tests are conducted on failed parts and the findings are the basis for formulating a test program for confirming defects in parts. The tests are:

(i) The failed parts are visually examined externally at 10X magnification or higher for physical damages such as cracks and chip-off, discoloration and loosening of terminations or other mechanical piece parts. Date codes on the parts are noted.

(ii)   Passive parts such as switches and variable resistors are operated to identify functional defects.

(iii) Electrical performance tests are conducted on failed parts as per relevant slash sheet and the general specification. If feasible, the voltages at the input/output pins are verified for ICs. Semiconductor devices are tested for short and open between terminals.

(iv)   One or two failed parts are opened for microscopic internal examination. Manufacturers recommended procedures are followed for de-capping ICs and semiconductor devices. For other passive components, the parts are disassembled after understanding their assembly procedures.

7.6.3 Test Program for ICs and Semiconductor Devices

Failure cause analysis on ICs and semiconductor devices is identifying defects in the design and manufacturing of the devices. The analysis requires engineering expertise in the technology of the devices and expensive facilities. Support of device manufacturers and well-equipped laboratories are essential to fix the failure causes of the devices, such as metallization or molding defects. Failure analysis on ICs and semiconductor devices is therefore limited to confirming the existence of defects in components using a test program and the findings are the basis to initiate root cause analysis by device manufacturers. Further, considerable time and efforts are required for conducting tests to confirm defects in ICs and semiconductor devices. Hence, it should be ensured that component failures are not induced by design errors or by manufacturing processes before suspecting the quality of the components.

Qualification and screening tests are specified in MIL-PRF-38535 for ICs and for semiconductor devices in MIL-PRF-19500. The purpose of the tests is explained in the Standards for Test Methods, MIL-STD-883 and MIL-STD-750. Test programs for ensuring the quality and reliability of ICs and semiconductor devices are published by manufacturers. The Quality and Reliability Handbook [9] of On Semiconductors, USA, lists the performance failure modes (Ex. parametric shifts and high leakage) that would be exposed by reliability stress tests on the devices. The findings on the examination and tests on failed parts, military standards, and the test programs of manufacturers are correlated to derive the test program for confirming defects in ICs and semiconductor devices from the tests listed below:

Integrated circuits (MIL-PRF-38535):

(i)   ESD testing for CMOS ICs
(ii)   Autoclave (for plastic packages)
(iii) Highly Accelerated Stress Testing (for plastic packages)
(iv)   Temperature cycling
(v)   Thermal shock

  (vi)   Soldering heat
 (vii)   Steady-state life test

   Semiconductor devices (MIL-PRF-19500):

    (i)   ESD testing for CMOS devices
   (ii)   High temperature non-operating life
  (iii)   Temperature cycling
  (iv)   High temperature reverse bias
    (v)   High temperature gate bias
   (vi)   Thermal shock
  (vii)   Temperature cycling
 (viii)   Intermittent operation life
   (ix)   High temperature life (non-operating)

7.6.4 Test Program for Passive Components

The method of deriving test program for confirming defects in passive components
is same as that of ICs and semiconductor devices except that the Standard for Test
Methods, MIL-STD-202 is used. Standard test equipment and facilities are adequate for conducting tests. Unlike ICs and semiconductor devices, the root causes
of failures could also be established for most of the passive components. The
findings on the examination and tests on failed parts, military standards, and the
test programs of manufacturers are used to derive the test program for confirming
defects in passive components from the tests listed below:

    (i)   Materials and finishes of piece parts
   (ii)   Critical electrical and mechanical tests
  (iii)   Life (electrical)
  (iv)   Life (Mechanical)
    (v)   Thermal shock
   (vi)   Vibration /Random vibration
  (vii)   Moisture resistance
 (viii)   Corrosion

7.7 Corrective Actions and Reporting

Having established the causes of component failures, corrective actions are
identified and implemented to prevent the recurrence of the failures. Corrective
actions could be design changes or improving manufacturing processes or vendor
development or their combination. Multiple corrective actions are identified

considering feasibility, time and cost for implementing the actions. A failure analysis report is prepared and the report contains:

  (i)   Description of failure
 (ii)   Confirmation of the failure
(iii)   Examination and tests on failed parts
(iv)   Approach for failure analysis
 (v)   Test program
(vi)   Failure causes with evidences
(vii)   Categorizing the origin of the causes as design error or manufacturing process induced error or defective components
(viii)   Recommended corrective actions
(ix)   Learning from the failure for manufacturing and new designs

Assessment Exercises

1. Say True or False:

  (i)   Discussing corrective actions in the brainstorming session for identifying the failure causes of component failures is useful to complete the failure analysis faster.
 (ii)   Defects in components are the prime reason for the failure of components in electronic equipment.
(iii)   Soldering and post-cleaning operations on PCBs could induce thermal shocks on electronic components.
(iv)   The requirements of a component (Ex. 2N3019) are mentioned in the general specification of military standard applicable for the component.
 (v)   Military slash sheet for a non-mil failed component should be procured for preparing test schedule for confirming defects in the failed component.

2. Describe the general approach for identifying the root causes of electronic component failures.
3. What are the sources of component failures?
4. Explain FDMS showing the inputs and the outputs of the system.
5. Explain the administrative requirements of a failure data management system.
6. Explain how the findings on the failure analysis of components could be used for new designs.

Research Assignments

1. Component and electrical performance failures might be observed when conducting reliability improvement tests on the bought-out products listed below. Identify the root causes of the failures and suggest at least three corrective actions to eliminate the causes of the failures:

   (i)   SMPS
   (ii) RF Power Amplifier
   (iii) LED Lighting unit
   (iv) Audio power amplifier
   (v)   Battery charger

References

1. Deppe, R.W., Minor, E.O.: Reliability enhancement testing (RET). In: IEEE Proceedings Annual Reliability and Maintainability Symposium, pp 91–98 (1994)
2. Failure Mechanisms of Semiconductor Devices, T0400BE-3, 2009.4, Panasonic Corporation, USA
3. MIL-STD-883G:2006, Test Method Standard for Microcircuits
4. MIL-STD-750-1:2013, Part-1, Environmental Test Methods for Semiconductor Devices
5. MIL-STD-202G:2002, Test Method Standard Electronic and Electrical Component Parts
6. Bot, Y.: improving product's reliability by stress de-rating and design rules check. In: IEEE Proceedings Annual Reliability and Maintainability Symposium, 978-1-4673-4711-2/13 (2013)
7. NEC Electronics Corporation, Japan: Optical/Microwave Semiconductor Devices: Review of Quality and Reliability Handbook, Document PQ10478EJ02V0TN, 2nd edn. (2006)
8. Natarajan, D.: A Practical Design of Lumped, Semi-lumped & Microwave Cavity Filters. Springer, Berlin (2013)
9. Quality and Reliability Handbook , HBD851/D, on Semiconductor, USA, Rev. 7 (2013)

Chapter 8
Reliability Database System for Design

Abstract Reliability database system (RDBS) provides assistance for integrating reliability and maintainability requirements concurrently with the design and development activities of electronic equipment. Designing the database system to provide support for the application of components, overstress analyses, FMEA, internal verification testing, qualification testing, and failure analysis reports is explained.

8.1 Reliability Database System

Reliability database is the collection of pertinent information from the catalogs of component manufacturers, the application notes of devices, military standards, and literature for the design and development of electronic equipment. It also reflects the internal experience of organizations. Reliability database system (RDBS) organizes and accesses the information using computer programs to derive the desired outputs. The database system provides assistance for integrating reliability and maintainability requirements concurrently with the design and development activities of electronic equipment. Concurrent design minimizes redesign efforts. The database system itself provides stability to product design organizations.

8.1.1 Capabilities

RDBS is unique to the needs of an organization. There is no limit to designing the capabilities of the database system to provide assistance for reliable design of equipment. Simple or automated software packages [1] can be developed for accessing reliability database to provide assistance for the implementation of the reliability activities. The capabilities that would be adequate for designing electronic equipment and COTS items are shown in Fig. 8.1 and the activities are:

© Springer International Publishing Switzerland 2015   99
D. Natarajan, *Reliable Design of Electronic Equipment*,
DOI 10.1007/978-3-319-09111-2_8

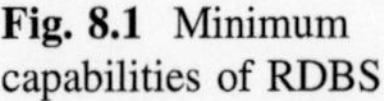

Fig. 8.1 Minimum
capabilities of RDBS

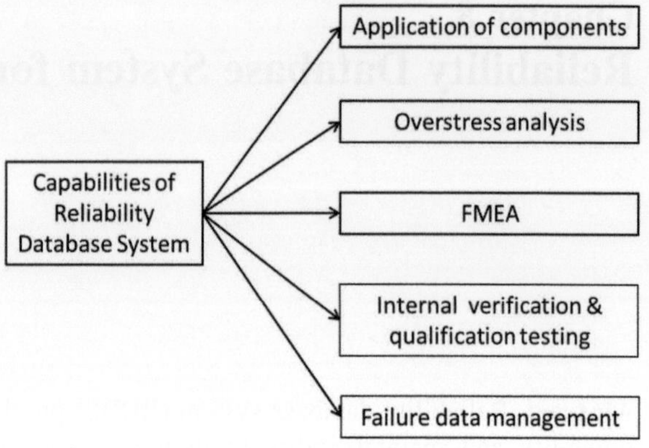

(i)   Application of components
(ii)   Overstress analyses
(iii) FMEA
(iv)   Internal verification and qualification testing
(v)   Failure analysis reports

8.1.2 Design and Development

RDBS is the means to ensure that the expertise of QA is made available virtually to circuit designers. Concurrent design becomes a way of life for designers in RDBS. The design and development of RDBS require focused and sustained efforts. An effective organization is required for designing, developing, and maintaining RDBS as per organizational needs. The responsibilities could be assigned to the internal QA team; the organization chart is shown in Fig. 8.2. The QA team must be committed to acquiring expertise and updating the RDBS continually. The commitment of QA is essential for the success of RDBS. Practical considerations in the design of RDBS are presented. Developing software packages is not explained as it is outside the scope of the presentation.

8.1.3 Practical Considerations

RDBS is a fairly large software system. Controls as per international standards, such as ISO9001, should be applied for the design and development of the database system. Procedures should be developed for controlling the process of design, development, testing, and documentation of the system. Discussions with circuit designers provide valuable inputs for designing the database system. The system has numerous records of data for providing assistance for reliable design of

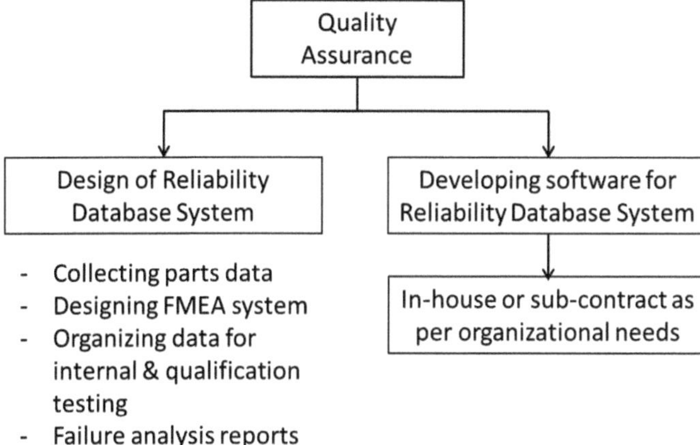

Fig. 8.2 Design and development of RDBS

equipment. Useful information for designing RDBS and controlling the records of data is:

(i)   RDBS should be designed to serve the design and development processes of equipment. Reliability administrative efforts such as documentation and multiple searches should be eliminated for designers.

(ii) Reliable application of ICs and semiconductor devices practically decides the reliability of equipment. Priority should be set for collecting data on the devices for current and near future projects.

(iii) The records of data should be authenticated by their sources with references. It is possible that some of the records of data cannot be authenticated for want of references. Such records should be marked as "suspect" or with suitable status identification. The status could be changed to normal as and when references become available.

(iv) Revision date should be indicated for the records of data. A new date is indicated when a record is amended. The obsolete records of data should also be preserved for knowledge purposes.

(v)   Some of the records of data might be required by many reliability activities. Duplication of records of data that are common to reliability activities should be avoided for controlling the revision of records.

8.2 Parts Database

The database of electronic components is the foundation of RDBS. Maximum ratings and application data are the elements of the database. Collection data requires focus and patience in understanding the components and their applications. The database provides assistance for:

(i) Application of components
(ii) Overstress analyses

Maximum ratings of components are available in the catalogs of manufacturers. Application information is distributed in the catalogs and application notes of competitors. Military standards and literature might also have to be referred. The elements of part database and their applications are explained for MOSFETs and multilayer ceramic chip capacitors.

8.2.1 Data Elements for MOSFETs

The data elements of MOSFETs are:

(i)   Absolute maximum ratings
(ii)   Application reliability information
(iii)   Optional information such as explaining characteristics of maximum ratings and the methods of computing power dissipation

8.2.1.1 Absolute Maximum Ratings

The absolute maximum ratings of MOSFET characteristics are shown in Table 8.1. The maximum values vary between MOSFETs and hence the values are shown as "XX". Data sheets of manufacturers might not specify all the characteristics shown in Table 8.1 or they might specify additional characteristics. Engineering decisions are used to finalize the list of characteristics. Thermal resistance and internal limit for junction temperature are also included in the list of characteristics for applying de-rating.

8.2.1.2 Application Reliability Information

The application information of MOSFET is shown in Table 8.2 with references. The references should be available in pdf format for retrieval when required by designers.

8.2.2 Application of MOSFET Data Elements

The application of MOSFET data elements is explained for Peak Diode Recovery dv/dt, and operating junction temperature, T_j. The maximum limits for the characteristics of MOSFET in SOT-223 package are assumed. The circuit design data as needed by the program of RDBS should be entered.

Table 8.1 Absolute ratings of MOSFET—example

Characteristic (unit)	Symbol	Max. limit	Conditions
Drain-to-source voltage	V_{DSS}	XX	–
Continuous drain current @ 25 °C	I_D	XX	–
Continuous drain current @ 70 °C	I_D	XX	–
Pulsed drain current (A)	I_{DM}	XX	Limited to T_{j-int}
Gate-to-source voltage (V)	V_{GS}	XX	–
Single pulsed avalanche energy (mJ)	E_{AS}	XX	Turn-off in unclamped inductive load circuits
Repetitive avalanche energy (mJ)	E_{AR}	XX	Limited to T_{j-int}
Repetitive avalanche current (A)	I_{AR}	XX	Limited to T_{j-int}
Peak diode recovery dv/dt (V/nS)	dv/dt	XX	For circuits using body drain diode
Total power dissipation at 25 °C ambient temperature (W)	P_D	XX	P_D reduces XX W/°C
Junction temperature (°C)	T_{j-max}	XX	–
Internal limit for T_j (°C)	T_{j-int}	XX	–
Lower temperature (°C)	T_{L-stg}	XX	–
Upper temperature (°C)	T_{U-stg}	XX	–
Thermal resistance, θ_{ja} (°C/W)	θ_{j-a}	XX	–
General de-rating (%)	–	10	Internal

Table 8.2 Application information for MOSFET—example

Application information	References
E_{AS} and the peak V_{DS} should be measured for unclamped inductive switching MOSFET circuit	Fairchild AN-9010, Rev. 1.0.5, Apri, 2013
Peak value of diode recovery dv/dt and the peak V_{DS} should be measured for MOSFET circuits that use the parasitic body drain diode	Fairchild AN-9010, Rev. 1.0.5, April 2013
Operating junction temperature of temperature of MOSFET could be reduced by 10–15 % by optimizing the design of copper mounting pads	Fairchild AN-1028, Rev. B, Augest 1998

Example-1 for peak diode recovery dv/dt:

- Maximum limits for MOSFET:

    - V_{DSS} (V): 100
    - Peak Diode Recovery dv/dt (V/nS): 6.0

- Circuit design data inputs by designers:

    - Supply voltage, V_{DD}, (V): 90
    - Parasitic body diode used
    - Peak Diode Recovery dv/dt (V/nS): 4.1

- Peak Drain-to-Source Voltage (V): 95

- System outputs:

  - Supply voltage, V_{DD}, (V): 90/ACCEPT
  - Peak Diode Recovery dv/dt (V/nS): 4.1/ACCEPT
  - Peak V_{DS} (V): 95/General de-rating not complied

Example-2 for operating junction temperature, T_j:

- Maximum limits for MOSFET:

  - P_D at T_A, 25 °C (W): 2.5
  - P_D reduces by 0.036 W/°C
  - T_{j-max} (°C): 150
  - T_{j-int} (°C): 130
  - θ_{j-a} (W/°C): 50

- Circuit design data inputs by designers:

  - Operating T_A (°C): 70
  - P_D (W): 1.4

- System outputs:

  - Operating T_j (°C): 140 /within T_{j-max} but T_{j-int} exceeded

Application information	References
Operating junction temperature of temperature of MOSFET could be reduced by 10–15 % by optimizing the design of copper mounting pads	Fairchild AN-1028, April 1996

Provision to access AN-1028 is a part of system output. Visual effects could be designed in the display of system outputs. The message, "ACCEPT" is displayed with a green background. Messages for exceeding internal de-rating limit are displayed with a yellow background. The message, "REJECT" is displayed with a red background.

8.2.3 Data Elements for Ceramic Chip Capacitors

Collecting part data on passive components is relatively simpler compared to ICs and semiconductor devices. The data elements of multilayer ceramic chip capacitors are:

(i)   Maximum ratings:

  - Rated voltage (V)
  - Lower operating temperature (°C)
  - Upper operating temperature (°C)
  - Case temperature for high power RF applications (°C)

Table 8.3 Application information for ceramic chip capacitor—example

Application information	References
In high power RF applications, the equivalent series resistance (ESR) is significant causing rise in the case temperature of chip capacitors. Selecting higher case size capacitors is a solution	(i) ATC #001-942 Rev. C; 4/05
	(ii) ATC #001-923 Rev. D; 4/07
By mounting the chip capacitor vertically, the first parallel resonance will not be present thereby significantly extending the usable bandpass	ATC #001-821 Rev. D; 10/05

(ii)   Internal de-rating limits:

- Internal de-rating for case temperature (%): XX
- Internal general de-rating (%): XX

(iii)   Application reliability information in Table 8.3:

8.2.4 Application of Chip Capacitor Data Elements

The application of multilayer ceramic chip capacitor data elements is explained for upper operating temperature. The rated temperature of the capacitor and the internal de-rating limit for the temperature are assumed. The circuit design data as needed by the program of RDBS should be entered.

- Maximum limits for chip capacitor:

    - Upper operating temperature (°C)125
    - Case temperature (°C)125
    - Internal de-rating limit for case temperature 20 %

- Circuit design data inputs by designers:

    - Operating T_A (°C)70
    - Operating frequency 1 GHz
    - Case temperature of capacitor (°C)105

- System outputs:

    - Operating T_A (°C): 70/ACCEPT
    - Case temperature of capacitor (°C): 105/Within T_{max} but T_{int} exceeded

Application information	Reference
In high power RF applications, the equivalent series resistance (ESR) is significant causing rise in the case temperature of chip capacitors. Larger case size capacitors could handle higher RF power	(i) ATC #001-942 Rev. C; 4/05
	(ii) ATC #001-923 Rev. D; 4/07

   Provision to access the application notes, ATC #001-942 and ATC #001-923 is a part of system output.

8.3 Database for FMEA

The requirements of FMEA software vary with organizational needs and the complexity of electronic equipment. Unlike part database, there are few data elements in the database of FMEA for use by designers. Most of the data elements are inputs from designers and the inputs are based on the technical analysis of circuits. Hence, understanding the technical requirements and providing operational comfort of circuit designers are the most important criteria for developing FMEA software. FMEA automation allows the analyst to concentrate more on the analysis than on the associated clerical tasks such as collecting and organizing input data, entering redundant data, and formatting reports [2]. The design of database and software should consider the benefits of FMEA (Sect. 4.4). Provision to monitor the status of implementing preventive actions by circuit designers is necessary for realizing the benefits of FMEA.

Libraries for failure modes, failure effects, failure causes, and preventive actions should be planned. The libraries should be classified based on frequency band, digital or analog signal processing, etc., for faster retrieval of data. Help menus should be planned liberally.

To begin with, the FMEA information generated by designers is stored in transit libraries and approved information is transferred to final libraries later. Both transit and final libraries are available to designers with suitable message for information from transit libraries. The concept of Fault Identification Number and grouping of equivalent failure modes would be useful in documenting and deriving the outputs of FMEA [3]. The FMEA automation software requirements applied by auto industries [4] could be considered in developing the software for electronic industries.

8.4 Database for Test Procedures

The database for preparing internal verification and qualification test procedures is in text format and is accessed by options and user inputs. Stand-alone environmental test procedures (Sect. 6.3.2) are prepared using the database. Thorough understanding of internal documents and the Standards for Tests Methods are required for designing the database system. Designing database and preparing stand-alone internal environmental test procedure are explained for high temperature test as per MIL-STD-810G [5]. Performance trend analysis (Sect. 6.3.4) could be integrated with the preparation of high temperature test procedure. For qualification testing, performance trend analysis is deleted in the environmental test procedure.

8.4.1 Database for High Temperature Test

The test method for high temperature test as per MIL-STD-810G contains:

(i)   Test process specifying test facility, controls on the test facility, preparation and test procedures for conducting high temperature test.
(ii) Guidance and information that are required for deciding test procedure and severities.

Guidance and information are really options for deciding high temperature test procedures from MIL-STD-810G. There are three procedures for the high temperature test and two of the procedures have two additional options, constant or cyclic exposure. The database for internal and qualification testing contains a total of five high temperature test procedures in text form. The database for high temperature test is simpler as one or two procedures are relevant for most electronic equipment and COTS items.

8.4.1.1 Accessing Database

The inputs for accessing the database for high temperature are derived from the contractual specifications of equipment and information from MIL-STD-810G. Examples of the inputs are:

(i) Select exposure type:

- Constant temperature storage
- Cyclic temperature storage
- Constant temperature operation
- Cyclic temperature operation
- Tactical-standby to operational

(ii) Assuming constant temperature operation is selected:

- Test temperature
- Location of thermocouple on the item under test for confirming temperature stabilization
- Enter list of pre-test, during test, and post-test performance parameters
- Enter list of in-circuit parameters for performance trend analysis

The output is the stand-alone high temperature test procedure in pdf format. The contents of the procedure are:

(i)   Objective in brief
(ii)   Referenced documents
(iii)   Details of test facilities with specifications
(iv)   Handling test interruptions
(v)   Test item description
(vi)   Test procedure steps with graphical representation

8.5 Database for Failure Analysis Reports

Database should be designed for component and electrical performance failures, categorizing the sources of failures as in Sect. 7.1.2. FMEA is for assumed failure modes, whereas failure analysis is for the failure modes that have occurred. Hence, FMEA format is suitable to design for recording failure analysis reports. The suggested database elements are:

- failure mode of module,
- details of failure,
- failure effects,
- root causes,
- origin of the root causes, (Sect. 7.3)
- corrective actions, and
- preventive actions,
- Failure analysis report in pdf format.

The preventive actions are information feedforward for new designs and they are linked to part database and FMEA as appropriate.

Research Assignments

1. Design and develop a computerized RDBS for the selection and application of components for the design of products listed below:

   (i)   SMPS
   (ii)   RF Power Amplifier
   (iii) LED Lighting unit
   (iv)   Audio power amplifier
   (v)   Battery charger

2. Design and develop a computerized system for handling the failure data observed during the reliability evaluation of the bought-out products listed below:

   (i)   SMPS
   (ii)   RF Power Amplifier
   (iii) LED Lighting unit
   (iv)   Audio power amplifier
   (v)   Battery charger

References

1. Arellano, L., Doshay, I.: Automating reliability prediction and de-rating analysis—an example of simultaneous engineering. In: IEEE Proceedings of the Annual Reliability and Maintainability Symposium, pp. 384–390 (1991)

2. Kukkal, P., Bowles, J.B., Bonnel, R.D.: Database design for failure modes and effects analysis. In: IEEE Proceedings of the Annual Reliability and Maintainability Symposium, pp. 231–239 (1993)
3. Spangler, C.S.: Equivalence relations within the failure mode and effects analysis. In: IEEE Proceedings of the Annual Reliability and Maintainability Symposium, pp. 352–357 (1999)
4. Montgomery, T.A., Pugh D.R., Leedham, S.T., Twitchett, S.R.: FMEA automation for the complete design process. In: IEEE Proceedings of the Annual Reliability and Maintainability Symposium, pp. 30–36 (1996)
5. MIL-STD-810G:2008: Test method standard environmental engineering considerations and laboratory tests

Chapter 9
Statistical Analysis of Field Failure Data

Abstract Cause analysis and implementing corrective actions for failures are fundamental requirements of analyzing field failure data. Additional engineering benefits could be realized by analyzing accumulated field failure data statistically. Analyzing the failure data of field replaceable electronic items using statistical probability distributions and predicting spares are illustrated with examples.

9.1 Introduction

Field failure data reported by the customers of electronic equipment is analyzed to identify the root causes of failures as explained in Chap. 7. Corrective actions are implemented to prevent the recurrence of the failures. Cause analysis and implementing corrective actions are the fundamental requirements of field failure data analysis. Additional engineering benefits could be realized by analyzing accumulated field failure data statistically.

Statistical analysis is presented in simple form so that engineers could feel at home with statistics. Introducing statistical distributions, illustrating the application of statistical distributions to estimate the failure rate of field replaceable items of electronic equipment, and predicting spares from the results of the analysis are presented. The statistical analysis should be continually updated with the deployment of additional equipment in field. The results of the statistical analysis would become more accurate with increase in the number of equipment deployed in field. The application of the techniques could be computerized but it is recommended to apply the techniques manually or using excel tool to begin with for deeper understanding of the techniques.

9.2 Statistical Distributions

Statistical distributions characterize the events that occur at random intervals. Examples of the events are accidents and failures. Mean and standard deviation are the most commonly used statistical measures that characterize the occurrence of

© Springer International Publishing Switzerland 2015

D. Natarajan, *Reliable Design of Electronic Equipment*,

DOI 10.1007/978-3-319-09111-2_9

events. Characterizing the random field failures using statistical distribution provides useful inputs for engineering and managerial decisions to improve the reliability and maintainability of electronic equipment. Three statistical distributions are relevant for analyzing the failure data of electronic equipment and they are:

  (i)   Poisson distribution
 (ii)   Exponential distribution
(iii)   Weibull distribution

9.2.1 Poisson Distribution

Poisson distribution is named after a French mathematician. The probability distribution of a Poisson random variable, r, is given by,

$$P(r) = \frac{e^{-\mu}\mu^r}{r!}$$

Poisson mean is μ and it is greater than zero. The standard deviation of Poisson distribution is $\sqrt{\mu}$. The number of failures in equipment is represented by the Poisson random variable, r. As the number of failures could only be integers (0, 1, 2, etc.), Poisson distribution is a discrete distribution.

Two applications of Poisson distribution is illustrated for analyzing field failure data of electronic equipment. Determining the failure rate (Poisson mean) of field replaceable PCBs is illustrated in Sect. 9.5 without showing the derivations for the Poisson mean. Any standard text book on statistics could be referred for the derivations. The second application of Poisson distribution is illustrated in Sect. 9.8 for predicting the spares of field replaceable PCBs that need to be stocked by manufacturers.

9.2.2 Exponential Distribution

The probability distribution function (P.D.F) of exponential random variable, t, is given by,

$$f(t) = \lambda e^{-\lambda t}$$
$$\text{Mean} = \text{Standard deviation} = 1/\lambda$$

Time to failures of field replaceable items electronic equipment is represented by the random variable, t. As time to failures could assume fractional values, exponential distribution is a continuous distribution. Cumulative distribution function (C.D.F) of exponential distribution is required for analyzing the failure data of field replaceable items as it directly gives the percentage of items that could be expected to fail for the random variable, t. C.D.F. is derived from the P.D.F and its application is illustrated in Sect. 9.7 for analyzing the field failure data of electronic items. The C.D.F. of exponential distribution is given by,

$$F(t) = 1 - e^{-\lambda t}$$

9.2.3 Weibull Distribution

Weibull distribution is named after a Swedish engineer and it could be used for analyzing failure data of both electronic and nonelectronic items. It is also a continuous distribution like exponential distribution. C.D.F of Weibull random variable, t, is given by,

$$F(t) = 1 - \exp\left[-\left(\frac{t}{\eta}\right)^{\beta}\right]$$

The Weibull distribution parameters are:

β: Shape parameter
η: Scale parameter or characteristic life

It is easier to estimate the Weibull parameters through graphical probability plotting of failure data, described in Sect. 9.6. The estimated value of the shape parameter, β, provides useful information.

$\beta = 1$:
The failure data could be analyzed using exponential distribution. Computation of mean life and assigning confidence intervals are simpler as explained in Sect. 9.7. Failure rate, λ, is given by $1/\eta$.

$\beta > 1$:
Failure rate ('hazard rate') is increasing and it shows the presence of wear-out failure mechanisms in the items under analysis. Mechanical components like ball bearings and nozzles experience β values in the range, 1.6–7.6, indicating old age wear-out [1].

$\beta < 1$:
Failure rate ('hazard rate') is decreasing. The reason could be that reliability growth testing on items is effective or the new design is better than the previous design.

9.3 Population of Item and Its Characteristics

Population of items and the characteristics of the items are explained for the COTS item, RF Oscillator, shown in Fig. 9.1. Population is the number of items for which field failure data is analyzed. Assume that 1,000 RF Oscillators are deployed in the field. The population size of each PCB of the RF Oscillator is also 1,000. If there are two Local DC Power Supply PCBs in the RF Oscillator as a redundant design, the population size of the power supply PCB would be 2,000. The characteristics of population should be independent and identically distributed for performing statistical analysis.

9.3.1 Independent

The characteristic, independent, requires that the failure of one PCB should not be caused by another PCB. In the RF Oscillator, the failures of all the PCBs except the Oscillator PCB are independent. The failure of the Buffer PCB might cause the failure of the Oscillator PCB. When failure data is received for the Oscillator PCB, it should be included for analysis after confirming that the failure is not caused by the failure of Buffer PCB, i.e., the failure of Oscillator PCB is independent.

9.3.2 Identically Distributed

The second characteristic is that the failure data of population should be identically distributed. Statistically, this implies that if population has two types of components each having different underlying statistical distribution, then the failure data of each component should be analyzed separately.

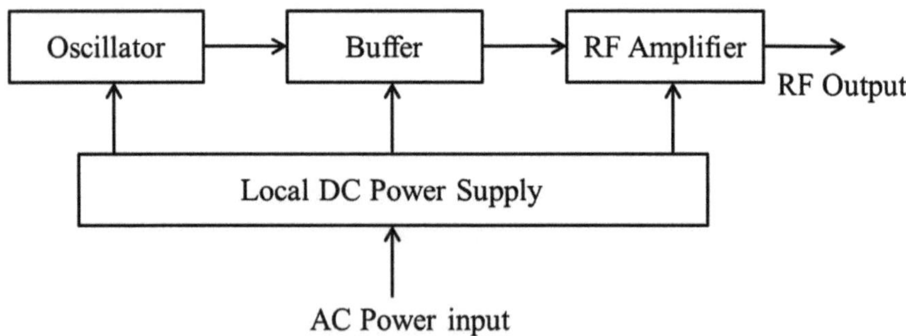

Fig. 9.1 The PCBs of RF oscillator

Assume that field failure data of an electronic unit having two modules needs to be analyzed. One module has only electronic components. The other module has electronic components and rotary coaxial joint. The failure data of the electronic unit should not be analyzed by combining the failure data from the modules as wear-out failure mechanism is associated with the module having rotary coaxial joint. The underlying statistical distributions of failure data from the modules are different. The failure data from the modules should be treated separately for statistical analysis. For the RF Oscillator, the failure data from the PCBs could be combined for analysis or they could be analyzed separately as the underlying statistical distribution of failure data from the PCBs is same.

9.4 Data Analysis Procedure

The procedure for analyzing field failure data is explained for the COTS item, RF Oscillator, shown in Fig. 9.1. It is applicable for electronic equipment also. The steps in analyzing the field failure data of RF Oscillator are explained, and the same are illustrated with a sample field failure data.

9.4.1 Define Field Replaceable PCBs

Although field failure data of RF Oscillator could be collected and analyzed as a single unit, it is more beneficial to collect data of the field replaceable PCBs of the RF Oscillator separately for analysis. Assume that the RF Oscillator is designed such that all the PCBs are field replaceable. The PCBs are:

(i)   Oscillator
(ii)   Buffer
(iii) RF Amplifier
(iv)   Local DC Power Supply

9.4.2 Obtain Time to Failure

Traceability procedure should be established linking the serial number of PCBs with the serial number of RF Oscillators before delivering to customers. This information is required to estimate time to failure of the PCBs when failures are reported from field by customers. Replacement details for failed PCBs should also be traceable for estimating the time to failure of the replaced PCBs. Time to failure should be collected for a minimum period of 1 year from the deployment of RF Oscillator. Estimating time to failure is illustrated with an example (Sect. 9.6.1).

9.4.3 Decide Statistical Distribution

Statistical distribution should be decided for analyzing time to failure data on the PCBs of RF Oscillator. Three solutions exist for deciding statistical distribution for analyzing field failure data. The simplest solution is to use the statistical distribution that has been applied successfully for analyzing field failure data of items having similar failure mechanisms [2]. Exponential distribution is found to fit for analyzing the field failure data of electronic items [3, 4]. It is relevant for analyzing the field failure data of the PCBs of the RF Oscillator.

The second method is to assume a statistical distribution function for the observed failure data and test the assumed hypothesis using a statistical test. Kolmogorov–Smirnoff test is commonly used for testing the hypothesis as it is powerful and completely nonparametric. Standard text books on statistics could be referred for hypothesis testing.

The third method is to use Weibull statistical distribution which is successfully used for analyzing fatigue failures, ball bearing failures, and many other items [5]. Weibull distribution could be used for analyzing the field failure data of electronic items. Exponential distribution is a special case of Weibull distribution.

9.4.4 Analyze Data

Laboratory reliability testing on electronic items could be designed in two ways. The test could be time terminated, i.e., the test is terminated after a pre-determined period. The test could be failure terminated, i.e., the test is continued until all the test items fail. Field failure data is treated as time terminated reliability testing and analyzed as per applicable statistical procedure. It is normally collected for a minimum period of 1 year from the date of deployment of electronic equipment in field.

Zero or more failures could be observed in the time terminated reliability testing of electronic items in field. The number of failures depends on the rigor of reliability program in the design and development of the electronic items. Failure data (time to failure) should be available for a minimum of ten electronic items to analyze the data with exponential or Weibull distribution [2]. If the number of failures is less than ten in the observation period, Poisson process is applied for analyzing field failure data. Application of Poisson process assumes constant failure rate, i.e., exponential distribution. The assumption is valid for the field failure data of electronic items.

Data analysis is presented for three cases of the RF Oscillator, shown in Fig. 9.1. The three cases are:

(i)   Application of Poisson process for Buffer PCB
(ii) Local DC Power Supply using Weibull
(iii) RF Amplifier using Exponential

9.5 Poisson Process for Data Analysis

Application of Poisson process for analyzing field failure data is illustrated with the Buffer PCB of the RF Oscillator. Total Test Time (T) of the Buffer PCB of the RF Oscillator in field should be estimated. Total Test Time is defined as the product of number of items deployed in field and the number of operating hours of the items in field. For example, if 50 Buffer PCBs are operating for 100 h in field,

$$\text{Total Test Time } (T) = 50 * 100 = 5{,}000 \text{ item-hours}$$

9.5.1 Estimating Total Test Time

Total Test Time (T) of Buffer PCB is same as that of RF Oscillator as one number of Buffer PCB is used in the oscillator. Estimating Total Test Time for the Buffer PCB is a cumulative process as the RF Oscillators would be delivered in batches to customers. Computerized traceability procedure as explained in Sect. 9.4.4 is used for estimating Total Test Time. The assumptions for estimating Total Test Time for the Buffer PCBs are:

(i)   A batch of 80 RF Oscillators is delivered to customers on the first day of every month.
(ii) RF Oscillators operate approximately for 350 h per month.
(iii) Failure data (time to failure) is collected and analyzed for 1 year from the date of delivery of the first batch of RF Oscillators

The monthly Item-Hours for the batches of RF Oscillators delivered in 1 year and the estimated Total Test Time are shown in Table 9.1.

Table 9.1 Estimated total test time for buffer PCB

Month	Quantity delivered	Months in field	Item-hours in 1 year = qty. * months * operating h/month
Jan	80	12	80 * 12 * 350 = 336,000
Feb	80	11	80 * 11 * 350 = 308,000
Mar	80	10	80 * 10 * 350 = 280,000
Apr	80	9	80 * 9 * 350 = 252,000
May	80	8	80 * 8 * 350 = 224,000
Jun	80	7	80 * 7 * 350 = 196,000
Jul	80	6	80 * 6 * 350 = 168,000
Aug	80	5	80 * 5 * 350 = 140,000
Sep	80	4	80 * 4 * 350 = 112,000
Oct	80	3	80 * 3 * 350 = 84,000
Nov	80	2	80 * 2 * 350 = 56,000
Dec	80	1	80 * 1 * 350 = 28,000
Total Test Time (T) for buffer PCB			2,184,000

9.5.2 Computing Failure Rate

Application of Poisson process for the Buffer PCB is illustrated with zero, and two failures in Total Test Time (T). Time to failure should be recorded for every failure as explained in Sect. 9.7.1 but it is not required for the application of Poisson process. Failure rate is normally expressed with 90 % confidence limit with lower and upper limits. Additional information is regarding confidence limit is available in Sect. 9.6.5.

$$\text{Upper limit of failure rate } (\lambda_U) = \frac{\chi^2_{\alpha/2,2(x+1)}}{2T}$$

$$\text{Lower limit of failure rate } (\lambda_L) = \frac{\chi^2_{(1-\frac{\alpha}{2}),2x}}{2T}$$

$\chi^2_{\alpha/2,2(x+1)}$ is the percentage point of the χ^2 (chi-square) distribution with $2(x + 1)$ degrees of freedom for the upper limit of failure rate (λ_U). The notation, α, represents the area (probability) to the right of the variable of the distribution. The notation, x, represents the number of failures observed in the Total Test Time (T). Statistical table is referred for obtaining the value of $\chi^2_{\alpha/2,2(x+1)}$. The same procedure is applicable for the lower limit of failure rate (λ_L) with $2x$ degrees of freedom.

The notation, v, is used to represent the degrees of freedom in statistical table.
$v = 2(x + 1)$
$\alpha = 1 - \Phi$ where Φ is the fractional value of confidence level
For 90 % confidence level, $\Phi = 90/100 = 0.9$
$\alpha = 1 - 0.9 = 0.1$
Statistical table might use the notation, P, in percentage instead of α.
$P = 100 - 90 = 10$ %

9.5.2.1 Case 1: Zero Failures

Zero failures are reported by customers in Total Test Time (T).
Total Test Time (T) = 2,184,000 item-hours
$x = 0$
$v = 2(0 + 1) = 2$
Confidence level = 90 %
$\Phi = 90/100 = 0.9$
$\alpha = 1 - 0.9 = 0.1$
$\chi^2_{\alpha/2,2(x+1)} = \chi^2_{0.05,2} = 5.991$ (from statistical table)

$$90\,\% \text{ upper confidence limit of failure rate } (\lambda_U) = \frac{\chi^2_{0.05,2}}{2T} = \frac{5.991}{2 * 2184000}$$
$$= 1.4\text{E-}06 \text{ failures}/\text{h}$$

90 % lower confidence limit for failure rate (λ_L) cannot be calculated as the degrees of freedom is zero. The maximum failure rate of the Buffer PCB of the RF Oscillator is 1.4E-06 with 90 % confidence level.

9.5.2.2 Case 2: Two Failures

Two failures are reported by customers in Total Test Time (T) and the failures are replaced with new Buffer PCBs.

Total Test Time $(T) = 2{,}184{,}000$ item-hours

Internal failure analysis has confirmed the existence of the two failures.

$x = 2$

$v = 2(2 + 1) = 6$

Confidence level $= 90\,\%$

$\Phi = 90/100 = 0.9$

$\alpha = 1 - 0.9 = 0.1$

For upper limit, $\chi^2_{\alpha/2,2(x+1)} = \chi^2_{0.05,6} = 12.592$ (from statistical table)

For lower limit, $\chi^2_{(1-\frac{\alpha}{2}),2x} = \chi^2_{0.95,6} = 1.635$ (from statistical table)

$$90\,\% \text{ upper confidence limit of failure rate } (\lambda_U) = \frac{\chi^2_{0.05,6}}{2T} = \frac{12.592}{2 * 2184000}$$
$$= 2.9\text{E-}06 \text{ failures}/\text{h}$$

$$90\,\% \text{ lower confidence limit of failurerate } (\lambda_L) = \frac{\chi^2_{0.95,6}}{2T} = \frac{1.635}{2 * 2184000}$$
$$= 0.4\text{E-}06 \text{ failures}/\text{h}$$

The failure rate of the Buffer PCB lies in the interval, (0.4–2.9) E-06 with 90 % confidence level.

9.6 Weibull Distribution for Data Analysis

Analyzing field failure data using Weibull distribution is illustrated for the Local DC Power Supply PCB. Time to failure should be recorded for analyzing the failure data for estimating the failure rate of the PCB. The assumptions for the analysis are:

(i)   A batch of 100 RF Oscillators is delivered to customers on the first day of every month.

(ii)   Serial number of the Local DC Power Supply PCB is the serial number of the RF Oscillator.

(iii) RF Oscillators operate approximately for 350 h per month.

(iv)   Failure data is collected and analyzed for 1 year from the date of delivery of the first batch of RF Oscillators.

(v)   Customer has reported ten failures of the Local DC Power Supply PCB in 1 year.

(vi)   The serial number of the failed PCB and the date of failure are available in the failure reports of customers.

9.6.1 Estimating Time to Failure

Traceability procedure linking the serial number and the source of PCBs with the serial number of RF Oscillators before delivery to customers should be computerized. When failures in Local DC Power Supply PCBs are reported by customers, the software tracks the serial number of the failed PCBs with the delivery database and indicates time to failure for the PCBs. Estimating time to failure is illustrated for one of the failure reported by customers. The same procedure is used to estimate time to failure for the remaining nine failures. The failure data is shown in Table 9.2.

Failure data as reported by customer: Sl. No. 126 failed on 22 Nov
Date of delivery for the PCB, 126, from internal database: 01 Jan
Days to failure: 326 days

Table 9.2 Estimated time to failure for local DC power supply

Month	Sl. nos. delivered	From customer report		Days to failure	Time to failure
		Sl. no.	Failure date		
Jan	101–200	126	22 Nov	326	3,803
Feb	201–300	202	18 Feb	18	210
Mar	301–400	360	2 Jun	94	1,097
Apr	401–500	441	3 Oct	186	2,170
May	501–600	569	24 Aug	116	1,353
Jun	601–700	690	3 Jul	33	385
Jul	701–800	790	17 Dec	170	1,983
Aug	801–900	851	4 Oct	65	758
Sep	901–1,000	–	–	–	–
Oct	1,001–1,100	1,010	29 Nov	60	700
Nov	1,101–1,200	1,150	10 Nov	10	117
Dec	1,201–1,300	–	–	–	–

Operating hours of the PCB per day $= 350/30$
Time to failure for the PCB, $126 = 326 * 350/30 \approx 3{,}803$ h

9.6.2 Weibull Probability Plotting

Weibull C.D.F is assumed for analyzing the failure data and estimating the failure rate (λ) of the Local DC Power Supply PCB. Statistical software packages like MINITAB could be used to validate the assumed Weibull distribution and estimate the Weibull parameters, β and η, by inputting time to failures. Alternatively, probability plotting, a graphical method, could be used. The graphical method is illustrated for the failure data of Local DC Power Supply PCB of the RF Oscillator.

9.6.2.1 Ordering Time to Failures

Time to failures is arranged in ascending order as shown in Table 9.3 for analyzing failure data. The serial numbers of the failed PCBs are no more required for the data analysis.

9.6.2.2 Compute $F(t)$

The underlying C.D.F. for the failure data is assumed to Weibull. Ten failures are tabulated in Table 9.3 and hence the sample size (n) of failure data is equal to ten. The Median rank values of $F(t)$ are computed using Bernard's approximation for sample sizes 10 or greater [6]. Bernard's approximation is:

$$F(t) = \frac{i - 0.3}{n + 0.4}$$

The computed values for $n = 10$ are shown in Table 9.4.

Table 9.3 Time to failures

Failure no. (i)	Time to failure (t) in hours
1	117
2	210
3	385
4	700
5	758
6	1,097
7	1,353
8	1,983
9	2,170
10	3,803

Table 9.4 Failure data with the values of $F(t)$

Failure no. (i)	Time to failure (t) in hours	$F(t) = \frac{i-0.3}{n+0.4}$
1	117	0.067
2	210	0.164
3	385	0.259
4	700	0.355
5	758	0.452
6	1,097	0.548
7	1,353	0.645
8	1,983	0.740
9	2,170	0.837
10	3,803	0.933

9.6.2.3 Validating the Assumed Weibull Distribution

Graphical plotting the C.D.F. of Weibull distribution with the values of time to failures (t) on x-axis and the values of $F(t)$ on y-axis is not useful for validating the assumed Weibull distribution. Hence, the C.D.F. is transformed to a linear equation, $y = mx + c$, using mathematical operations. Transformed values are obtained for both time to failures (t) and $F(t)$. Graphical plotting is done with the transformed values of t in x-axis and $F(t)$ in y-axis. If a straight line is obtained for the transformed values in the plotting, it confirms that the underlying C.D.F. is Weibull distribution. The method of transforming the C.D.F. of Weibull distribution into a linear equation is explained. The Weibull C.D.F. is given by,

$$F(t) = 1 - \exp\left[-\left(\frac{t}{\eta}\right)^{\beta}\right]$$

The expression is re-arranged as,

$$\frac{1}{1 - F(t)} = \exp\left[\left(\frac{t}{\eta}\right)^{\beta}\right]$$

Taking natural logarithm for both sides of the equation,

$$\ln\left[\frac{1}{1 - F(t)}\right] = \ln\left\{\exp\left[\left(\frac{t}{\eta}\right)^{\beta}\right]\right\}$$

$$\ln\left[\frac{1}{1 - F(t)}\right] = \left(\frac{t}{\eta}\right)^{\beta}$$

Fig. 9.2 Probability plotting
for the assumed Weibull

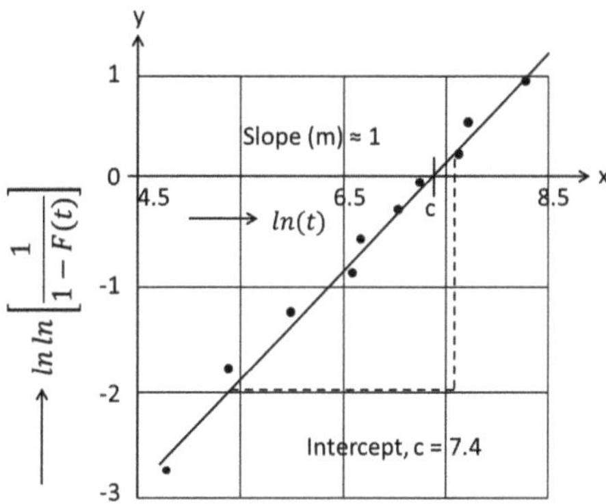

Taking natural logarithm again for both sides of the equation,

$$\ln\ln\left[\frac{1}{1-F(t)}\right] = \beta \ln(t) - \beta \ln(\eta)$$

The C.D.F. is transformed into a linear equation in the form, $y = mx - c$
The x-axis values are the logarithm of time to failures i.e. $\ln(t)$.

The y-axis values are the values of the expression, $\ln\ln\left[\frac{1}{1-F(t)}\right]$

The slope of the line, $m = \beta$
The intercept of the line on y-axis, $c = \beta \ln(\eta)$
The transformed values of x-axis and y-axis are shown in Table 9.5.
    The probability plotting for the assumed Weibull C.D.F. is shown in Fig. 9.2.
As the data fit into a straight line, the assumed Weibull C.D.F. for analyzing the
field failure data of the Local DC Power Supply is validated.

9.6.3 *Estimating Weibull Parameters*

The shape parameter (β) and the scale parameter (η) of the Weibull distribution are
estimated from the probability plotting, shown in Fig. 9.2.
    The slope of the line, $m = 1 = \beta$
    The intercept on x-axis, $c = 7.4 = \beta \ln(\eta)$
    $\eta = \exp(7.4/1) = 1{,}636$ h

λ, failure rate of Local DC Power Supply $\quad = (1/\eta) = (1/1{,}636)$
$$= 611\text{E-}06 \text{ failures/h}$$

    As the value of the shape parameter, β, is one, the failure data could be ana-
lyzed using exponential C.D.F. also.

Table 9.5 C.D.F. with transformed values of x-axis and y-axis

Failure no. (i)	Time to failure (t) (h)	$F(t) = \frac{i-0.3}{n+0.4}$	x-axis values, $\ln(t)$	y-axis values, $\ln\ln\left[\frac{1}{1-F(t)}\right]$
1	117	0.067	4.759	−2.707
2	210	0.164	5.347	−1.737
3	385	0.259	5.953	−1.204
4	700	0.355	6.551	−0.826
5	758	0.452	6.631	−0.509
6	1,097	0.548	7.000	−0.232
7	1,353	0.645	7.210	0.032
8	1,983	0.740	7.593	0.302
9	2,170	0.837	7.682	0.602
10	3,803	0.933	8.244	0.994

9.6.4 Weibull Probability Plotting Chart

Linear scale is used for the Weibull probability plotting shown in Fig. 9.2 and the points are plotted using transformed values. Weibull probability plotting chart (Ex.: Chartwell-6572) is available for direct plotting the values of time to failures (t) and $F(t)$. Using Weibull chart eliminates the need for transforming the values and it saves considerable efforts for analysis.

Weibull chart has transformed values for x-axis and y-axis. X-axis has logarithmic value of time to failure(t) and Y-axis is calibrated for the double logarithmic value of $[1/(1 - F(t)]$ in percentage. Time to failures (t) in hours and $F(t)$ in percentage could be directly entered. The chart has η-estimator horizontal dashed line. The value of η is given by the time to failure, where the η-estimator line cuts the straight line of the plotted data. The chart has estimation point and β-scale. A perpendicular is drawn from the estimation point to the straight line of the plotted data. The value of β is read from the β-scale where the perpendicular line cuts the β-scale.

9.6.5 Confidence Limits

The failure rate of the Local DC Power supply PCB, obtained from the Fig. 9.2, is called point estimate. The point estimate could be used for determining spares and for other managerial decisions provided the number of failure data points is 100 or more. As the number of failure data points is only ten, confidence limits should be assigned to the point estimate. Applying confidence limits provides interval estimate for failure rate. Assigning the limits is explained for 90 % confidence level.

Statistically, it means that the failure rate of an item lies within the estimated interval with 90 % confidence. Graphical method of assigning the limits is explained as it eliminates complex statistical computations.

9.6.5.1 Rank Values

The Weibull plotting in Fig. 9.2 uses Bernard's median rank values for $F(t)$ as explained in Sect. 9.6.2.2. For drawing the confidence limit curves for the plot, 10 and 90 % rank values for the sample size, 10, are required and they are obtained from Rank Tables or by using Rank calculators. The values of the ranks in percentage are shown in Table 9.6.

An improvised Weibull probability plotting chart, drawn to scale, is shown in Fig. 9.3 for drawing the Weibull lower and upper confidence lines. β-estimation point and β-scale are not shown in the chart. The η-estimator line is shown in dashed line. The values of time to failures (t) and median rank $F(t)$ are plotted on the improvised Weibull probability plotting chart in Fig. 9.3. The η-estimator line and the Weibull plot intersects at 1,630 h, which is the point estimate of η. The procedure to draw lower and upper confidence lines are explained for two points each. The points are selected such that one point is above the η-estimator line and the other point is below the η-estimator line.

9.6.5.2 Lower Confidence Limit

The value of 10 % rank for the sixth data point is 35.4 % and the median rank value is 54.8 %. A line is drawn horizontally from 35.4 % marking on the y-axis until it intersects the Weibull plot for the ten data points. An upward line as shown

Table 9.6 Values ranks for 90 % confidence limits

Failure no. (i)	Time to failures (t) (h)	$F(t)$, 10 % rank (%)	Median rank (%) $F(t) = \frac{i-0.3}{n+0.4}$	$F(t)$ 90 % rank (%)
1	117	1.1	6.7	20.6
2	210	5.5	16.4	33.7
3	385	11.6	25.9	45.0
4	700	18.8	35.5	55.2
5	758	26.7	45.2	64.6
6	1,097	35.4	54.8	73.3
7	1,353	44.8	64.5	81.2
8	1,983	55.0	74.0	88.4
9	2,170	66.3	83.7	94.5
10	3,803	79.4	93.3	99.0

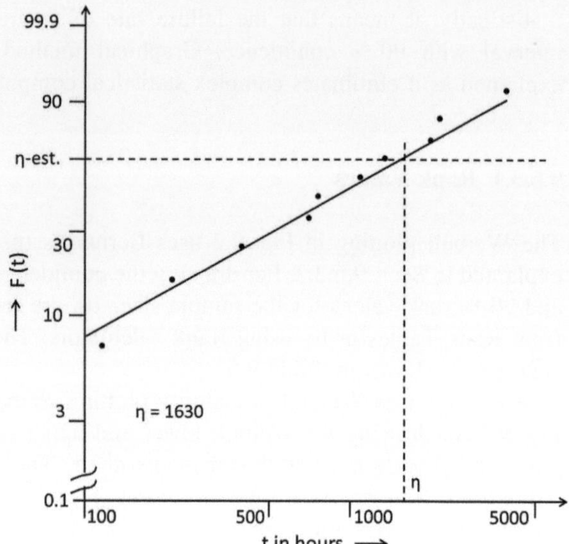

Fig. 9.3 Improvised Weibull probability plot with data points

by the arrow is drawn from the intersection point at until the median rank value, 54.8 % is reached. This is one point on the lower confidence line. Both the horizontal and the upward vertical arrow lines are shown in Fig. 9.4. The procedure is repeated for the value of 10 % rank for the eighth data point (55.0 %) and the median rank value (74.0 %) to obtain the second point on the lower confidence line. The two points are joined to obtain the lower confidence line for the Weibull plot as shown in Fig. 9.4. The η-estimator line intersects the lower confidence line at 900 h, which is the lower confidence level estimate of η i.e. η_L.

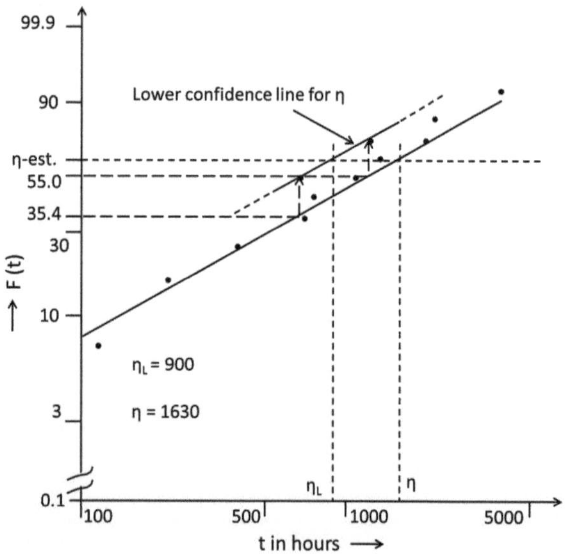

Fig. 9.4 Lower confidence limit for Weibull plot

9.6.5.3 Upper Confidence Limit

The value of 90 % rank for the sixth data point is 73.3 % and the median rank value is 54.8 %. A line is drawn horizontally from 73.3 % marking on the y-axis until it intersects the Weibull plot for the ten data points. A downward line as shown by the arrow is drawn from the intersection point at until the median rank value, 54.8 % is reached. This is one point on the upper confidence line. Both the horizontal and the vertical downward arrow lines are shown in Fig. 9.5. The procedure is repeated for the value of 90 % rank for the eighth data point (88.4 %) and the median rank value (74.0 %) to obtain the second point on the upper confidence line. The two points are joined to obtain the lower confidence line for the Weibull plot as shown in Fig. 9.5. The η-estimator line intersects the upper confidence line at 2,800 h, which is the upper confidence level estimate of η i.e. η_U. Both lower and upper confidence lines for the Weibull plot are shown in Fig. 9.6.

Failure rate (point estimate) $= 1/\eta = 1/1{,}630 = 613\text{E-}06$ failures/h

$$90\,\%\ \text{lower confidence limit for failure rate } (\lambda_L) = 1/\eta_U = 1/2800$$
$$= 357\text{E-}06 \text{ failures/h}$$

$$90\,\%\ \text{upper confidence limit for failure rate } (\lambda_U) = 1/\eta_L = 1/900$$
$$= 1111\text{E-}06 \text{ failures/h}$$

The failure rate of the Local DC Power Supply PCB of the RF Oscillator lies in the interval, $(357{-}1{,}111)\text{E-}06$ with 90 % confidence level.

Fig. 9.5 Upper confidence limit for Weibull plot

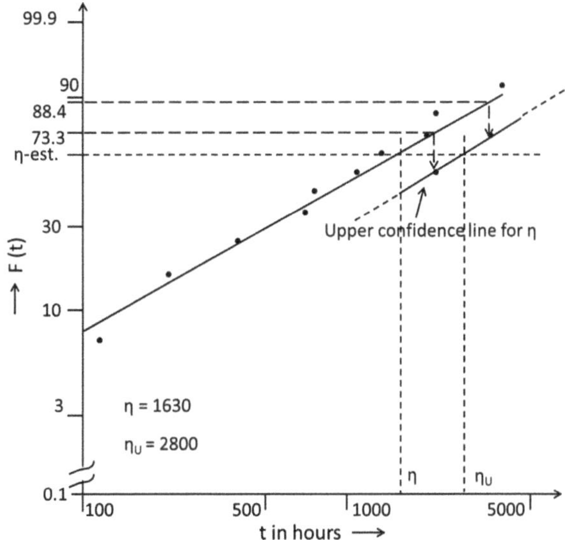

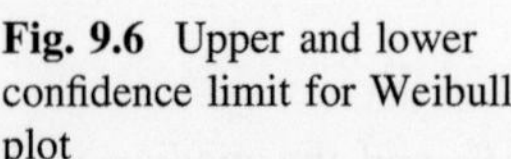

Fig. 9.6 Upper and lower confidence limit for Weibull plot

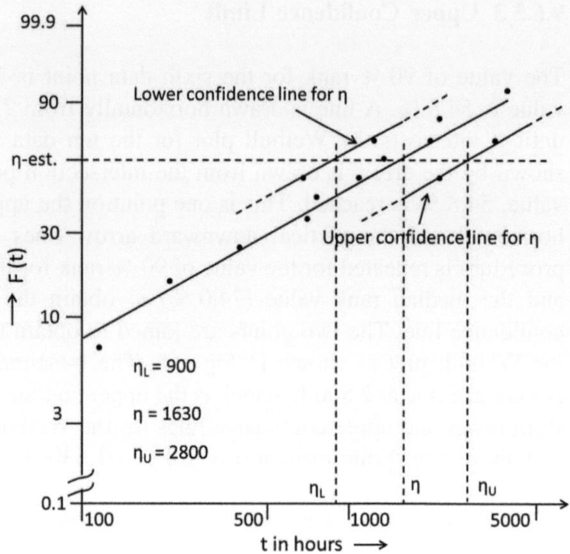

9.7 Exponential Distribution for Data Analysis

Analyzing field failure data using exponential distribution is illustrated for the RF Amplifier PCB of the RF Oscillator. Time to failures of the PCB is computed using the same assumptions and the procedure explained for the DC Power Supply PCB (Sect. 9.7). The ordered time to failures (t) and Bernard's approximation for $F(t)$ are shown in Table 9.7.

Table 9.7 Failure data with the values of $F(t)$

Failure no. (i)	Time to failure (t) in hours	$F(t) = \frac{i-0.3}{n+0.4}$
1	95	0.067
2	240	0.164
3	403	0.259
4	588	0.355
5	758	0.452
6	804	0.548
7	1,061	0.645
8	1,380	0.740
9	1,800	0.837
10	3,600	0.933

9.7.1 Exponential Probability Plotting

Graphical plotting the C.D.F. of exponential distribution is same as that of the Weibull distribution. The method of transforming the C.D.F. of exponential distribution into a linear equation is explained. The C.D.F. is given by,

$$F(t) = 1 - \exp(-\lambda t)$$

The expression is re-arranged as,

$$\frac{1}{1 - F(t)} = \exp(\lambda t)$$

Taking natural logarithm for both sides of the equation,

$$\ln\left[\frac{1}{1 - F(t)}\right] = \lambda t$$

The C.D.F. is transformed into a linear equation in the form, $y = mt$. The transformed values of y-axis are shown in Table 9.8. The x-axis values are time to failures (t). The y-axis values are the values of the expression, $\ln\left[\frac{1}{1-F(t)}\right]$.

The probability plotting for the assumed exponential C.D.F. is shown in Fig. 9.7. As the data fit into a straight line, the assumed exponential C.D.F. for analyzing the field failure data of the RF Amplifier is validated.

The slope of the line, $m = \lambda$

λ, failure rate of RF Amplifier $= 0.00076 = 760\text{E-}06$ failures/h.

Table 9.8 C.D.F. with transformed values for y-axis

Failure no. (i)	x-axis: time to failure(t) (h)	$F(t) = \frac{i-0.3}{n+0.4}$	y-axis values, $\ln\left[\frac{1}{1-F(t)}\right]$
1	95	0.067	0.070
2	240	0.164	0.178
3	403	0.259	0.301
4	588	0.355	0.440
5	804	0.452	0.601
6	1,061	0.548	0.794
7	1,380	0.645	1.033
8	1,800	0.740	1.349
9	2,417	0.837	1.811
10	3,600	0.933	2.698

Fig. 9.7 Probability plotting for the assumed exponential

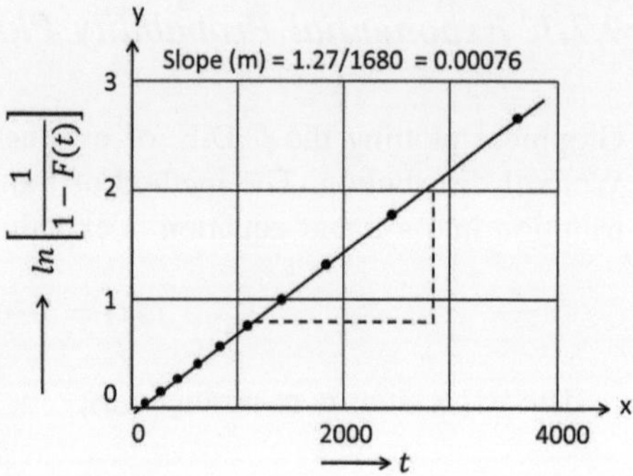

9.7.1.1 Exponential Probability Plotting Chart

Exponential probability plotting chart is available for direct plotting the values of time to failures (t) and $F(t)$. X-axis linear scale for time to failures (t) and y-axis is calibrated for the logarithmic value of $[1/(1 - F(t)]$ in percentage. Time to failures (t) in hours and $F(t)$ in percentage could be directly entered. The value of time to Failure (t) corresponding to 63 % of F(t) is the mean of the exponential distribution.

Mean $= 1/\lambda$

Failure rate $= 1/$Mean

9.7.2 Confidence Limits

The failure rate of the RF Amplifier, obtained from the Fig. 9.7, is the point estimate. As the number of failure data points is only ten, confidence limits should be assigned to the point estimate. Assigning the limits is explained for 90 % confidence level. Graphical method explained for the Weibull plot could be used but it is simpler to use statistical computations for obtaining the lower and upper confidence values for exponential distribution.

$$\text{Lower confidence limit for failure rate } (\lambda_L) = \frac{\chi^2_{(1-\alpha/2),\,2n}}{2\sum\limits_{i=0}^{n} t_i}$$

$$\text{Upper confidence limit for failure rate } (\lambda_U) = \frac{\chi^2_{\alpha/2,\,2n}}{2\sum\limits_{i=0}^{n} t_i}$$

$n = 10$

$v = 2 * n = 2 * 10 = 20$

Confidence level $= 90$ %

$\Phi = 90/100 = 0.9$

$\alpha = 1 - 0.9 = 0.1$

$$2\sum_{i=0}^{n} t_i = 2(95 + 240 + 403 + 588 + 804 + 1061 + 1380 + 1800 + 2417 + 3600)$$

$$= 24776$$

For lower limit, $\chi^2_{(1-\frac{\alpha}{2}),2n} = \chi^2_{0.95,20} = 10.851$ (from statistical table)

For upper limit, $\chi^2_{\alpha/2,2n)} = \chi^2_{0.05,20} = 31.41$ (from statistical table)

$$90\,\% \text{ lower confidence limit of failurerate } (\lambda_{\mathrm{L}}) = \frac{\chi^2_{0.95,20}}{2\sum_{i=0}^{n} t_i}$$

$$= \frac{10.851}{24776} = 438\text{E-06failures/h}$$

$$90\,\% \text{ upper confidence limit of failurerate } (\lambda_{\mathrm{U}}) = \frac{\chi^2_{0.05,20}}{2\sum_{i=0}^{n} t_i}$$

$$= \frac{31.41}{247761} = 1268\text{E-06failures/h}$$

The failure rate of the RF Amplifier PCB of the RF Oscillator lies in the interval, (438–1,268)E-06 with 90 % confidence level. The statistical computational method could also be used for assigning confidence limits for the point estimate of failure rate using Weibull C.D.F. provided the value of shape parameter, β, is close to unity.

9.8 Spares Prediction

Manufacturers maintain stocks of field replaceable items for supplying to customers when needed. The quantity of spares for the replaceable items could be statistically predicted using the estimated failure rate of field replaceable items [7]. Spares prediction procedure is explained and illustrated with examples. Additional benefits such as estimating warranty costs [3, 8], assessing contractual reliability requirements, and evaluating the reliability of designs [1] could also be obtained from the data analysis.

9.8.1 Determining Quantity of Spares

Determining the quantity of spares for electronic equipment is explained assuming that field servicing is restricted to the removal of defective PCBs or modules recommended by manufacturers and replacing them with identical new items. The defective items are nonrepairable in field with present advancements in technology and components. Poisson distribution is used for determining the quantity of spares. The method is accurate as applying exponential distribution for field failure data is valid for the field replaceable items of electronic equipment.

The probability distribution of Poisson random variable is,

$$P(r) = \frac{e^{-\mu}\mu^r}{r!}$$

One of the applications of Poisson distribution describes the occurrence of random events such as accidents in a time interval and μ could be replaced by λt [5]. Failures of electronic assemblies in field are also examples of random events. The probability distribution of Poisson random variable becomes,

$$P(r,t) = \frac{e^{-\lambda t}(\lambda t)^r}{r!}$$

λt represents the demand rate, which is the number of spares required to provide replacements for field failures. The demand rate, λt, is proportional to the number of items in field, operating hours per day, and failure rate of item [9]. Spares replenishing period is also a factor of the demand rate. Replenishing period is the period needed to manufacture spares for maintaining the recommended level of the spares. The demand rate, λt, should be modified reflecting the factors.

Determining the spares is illustrated for Buffer PCB of the RF Oscillator. The notations for determining the spares of the PCBs are:

n: Quantity of PCBs in RF Oscillator
N: Number of RF Oscillators deployed in field
t: Operating hours of RF Oscillator per day
λ_U: Upper limit failure rate of PCB in failures per million hours
T: Spares replenishing period in days
m: Demand rate
P: Probability of spares being available, when needed
x: Number of spares to be stocked by the manufacturer of RF Oscillator
m: demand rate $= n * N * t * \lambda_U * T * 1\text{E-}06$
The expression for computing spares is:

$$\sum_{u=0}^{x} \frac{e^{-m}m^u}{u!} \geq P$$

The least value of x, which satisfies the inequality, is the number of spares to be stocked by the manufacturer of RF Oscillator.

9.8.1.1 Spares for Buffer PCB

The values of the variables of Buffer PCB are:

n, Quantity of Buffer PCB in RF Oscillator $= 1$
N, Number of RF Oscillators deployed in field $= 1{,}000$ (assumed)
t, Operating hours of RF Oscillator per day $= 30$ h (assumed)
λ_U, Upper limit failure rate in failures/million h $= 2.9$ (from Sect. 9.5.2.2)
T, Spares replenishing period $= 5$ days (assumed)
m, Demand rate $= 1 * 1000 * 30 * 2.9 * 5 * 1\text{E-}06 = 0.44$
P, Probability of spares being available, when needed $= 0.9$ (assumed)

$$\sum_{u=0}^{x} \frac{e^{-0.44}0.44^u}{u!} \geq 0.9$$

x, Number of spares for Buffer PCB $= 1$

The number of predicted spares might be zero or very high. Engineering decisions could be made for deciding the stock quantity of spares as statistical analysis is an input for scientific decisions.

Assessment Exercises

1. Say True or False:

   (i) The occurrence of random failures in electronic equipment cannot be characterized.
   (ii) The failure rate of highly reliable electronic items with no failures in field operating conditions cannot be estimated.
   (iii) Exponential distribution which is characterized by constant failure rate is an example of discrete distribution.
   (iv) Weibull distribution could be used when the underlying statistical distribution of failure data is not known.

2. Explain population of item and its characteristics for analyzing failure data of the item.
3. Briefly explain the procedure for analyzing field failure data of electronic item.
4. The analysis of the field failure data of Frequency Modulator module, reported by customers has recorded time to failure (hours) as: 15029, 3484, 21985, 90559, 8924, 39712, 134924, 67428, 30067, and 51674. Estimate the interval estimate of failure rate at 90 % confidence level and the quantity of spare modules that should be stocked by manufacturer. State the assumptions used in the statistical analysis. Rank and statistical tables are provided.

References

1. Wang, W., Langanke, D.R.: Comparing two designs when the new design has few or no failures, is the new design better than previous one? In: IEEE Proceedings of Annual Reliability and Maintainability Symposium. pp. 322–325 (2001)
2. NIST: Engineering Statistics Handbook. NIST, Gaithersburg, June 2012
3. Lakey, M.J.: Statistical analysis of field data for aircraft warranties. In: IEEE Proceedings of Annual Reliability and Maintainability Symposium. pp. 340–344 (1991)
4. Tashtoush, G.M., Tashtoush, K.K., Al-Muhtaseb, M.A., Mayyas, A.T.: Reliability analysis of car maintenance scheduling and performance. Jordan J. Mech. Ind. Eng. 3(3), 388–393 (2010)
5. Chatfield, C.: Statistics for Technology, a Course in Applied Statistics. Chapman and Hall, New Jersey (1983)
6. Newton, D.W.: Course Tutor. Lecture Notes (1983–1984), University of Birmingham
7. Dreyer, S.L., Smith II, G.: Innovations in military spares analysis. In: IEEE Proceedings of Annual Reliability and Maintainability Symposium. pp. 397–401 (1995)
8. Park, M., Pham, H.: A new warranty policy with failure times and warranty servicing times. IEEE Trans. Reliab. 61(3) 822–831 (2012)
9. Bian, J., Guo, L., Yang, Y., Wang, N.: 2013, Optimizing spare parts inventory for time-varying task. Chem. Eng. Trans. 33, 637–642. doi: 10.3303/CET1333107

Appendix A
Reliability Testing for Non-standard Loads

A.1 Introduction

Adequate information is available in the catalogs and application notes of switches and electromechanical relays for switching standard loads (Sects. 3.3.7.2 and 3.3.8.2). Limited information is available for switching nonstandard loads in the application notes. The limited information is extrapolated for switching nonstandard loads as per circuit needs. The extrapolation might lead to uncertainties in the application reliability of the components. Hence, it is necessary to evaluate the reliability of components for switching nonstandard loads.

There are two methods for handling the uncertainties in switching nonstandard loads. Reliability guidance could be obtained from manufacturers, describing the exact application of switches and electromechanical relays in electronic circuits. Alternatively, reliability testing could be conducted to confirm the satisfactory application of the components. The application reliability testing of the components for nonstandard loads is explained.

A.2 Basis of Application Reliability Testing

Military Standards are available for the various categories of switches and electromechanical relays. The standards specify electrical life test for the contacts of switches and relays with various types of loads. The life test could be modified to evaluate the application reliability of the components for nonstandard loads. Designing application reliability test is explained for hermetically sealed relays and it is applicable for all types of switches and relays.

A.3 Application Reliability Problem:Example

Consider a hermetically sealed electromechanical relay with contact rated at 28 V/5 A DC resistive as per data sheet. The circuit application of the relay requires switching a resistive load of 48 V/1 A DC at the worst-case operating ambient

© Springer International Publishing Switzerland 2015
D. Natarajan, *Reliable Design of Electronic Equipment*,
DOI 10.1007/978-3-319-09111-2

temperature of 80 °C. The reliability of the application of the relay needs to be evaluated. Let the relay be denoted as RL_1. It is assumed that the relay is qualified to MIL-PRF-39016, which specifies the requirements of electromechanical hermetically sealed relays with contact ratings up to 5 A. If military slash sheet does not exist for a component, the procedure described in Sect. 7.6.1 could be followed for preparing the slash sheet.

A.4 Design of Application Reliability Test

The general specification, MIL-PRF-39016F, the MIL slash sheet of the relay, RL_1, and the manufacturer's data sheet are collected for designing the application reliability test. The military standard specifies electrical life test as part of qualifying the relays to the standard. The life test is specified for resistive, inductive, and lamp loads. Lamp load is relevant for circuits having high inrush current. The qualification requirements specified in the military standard for the life test are:

(i)   Initial electrical tests with acceptance criteria
(ii) Test conditions:

  - Test temperature
  - Load conditions
  - Number of cycles
  - Cycling rate
  - Duty cycle

(iii) Final electrical tests with acceptance criteria

The qualification requirements of the life test are modified to evaluate the application reliability of the relay, RL1, for nonstandard loads. The modifications are acceptable as the objective of the evaluation is to generate reliability information for deciding the satisfactory application of the relay.

A.5 Test Conditions

The recommended electrical tests and the life test conditions for evaluating the application reliability of the hermetically sealed relay, RL_1, are listed below. The tests are conducted as per the test methods described in MIL-PRF-39016F.

(i)   Number of relays for evaluation: 4
(ii) Initial electrical test:
      Initial electrical contact resistance is measured. The measured values should be within the limits specified in the military standard.
(iii) Test temperature:

The worst-case operating temperature for the relay, RL_1, is 80 °C. The military standard specifies 125 °C for the life test. Considering the objective of the evaluation, the test temperature could be fixed at 10–15 °C higher than the worst-case operating temperature of the relay.

(iv) Load conditions:
The switching load conditions are fixed as 48 V/1 A DC resistive.

 (v) Number of cycles:
The military standard specifies 100,000 on–off cycles for switching the load. The same could be applied.

(vi) Cycling rate:
The military standard specifies (20 ± 2) cycles/min for switching the load conditions. The same could be applied.

(vii) Duty cycle:
The military standard specifies equal on and off periods for resistive load. The same could be applied.

The relays are subjected to electrical life test as per MIL-PRF-39016, applying the test conditions.

A.6 Acceptance Test Criteria

Final electrical contact resistance is measured. The military standard specifies that the measured values shall be no greater than twice the initial specified contact resistance requirement for qualification. For example, if the initial specified limit is 50 mΩ maximum, the final measured values of contact resistance should be within 100 mΩ. As the application reliability of the relay is being evaluated for nonstandard load, it would be appropriate to compute the percentage change for the final measured values of contact resistance from the initial measured values and develop criteria for acceptance. The acceptable percentage change is decided by engineering considerations. A nominal change of 50 % could be considered.

Appendix B
Reliability Growth Testing

B.1 Definition

Reliability growth testing is one where testing is often used to uncover and correct problem failure modes during a series of test phases [1]. Series of tests are conducted on the proto-models of electronic modules and equipment for revealing latent performance and component failures. Conducting the tests on the proto-models is termed as reliability growth testing.

B.2 Reliability Growth

All the tests that are conducted on the proto-models of electronic modules and equipment are reliability growth tests. The reliability of the proto-models improves, i.e., grows after every test on the proto-models provided,

- failures in the tests are recorded,
- root causes are identified for the failures,
- corrective actions are implemented to eliminate the root causes, and
- effectiveness of the corrective actions is monitored.
- If corrective actions are found effective, reliability growth has occurred. If the actions are found ineffective, the process of analyzing for root causes, identifying corrective actions, and monitoring the effectiveness of the corrective actions is iterated. The reliability growth process is shown in Fig. B.1.

B.3 Reliability Growth Tests

Overstress analyses, performance trend analysis, thermal shock, random vibration, accelerated burn-in, and environmental tests are reliability growth tests that are conducted on the proto-models of electronic modules and equipment. Assuming that the reliability growth process, shown in Fig. B.1, is applied for the failures

© Springer International Publishing Switzerland 2015
D. Natarajan, *Reliable Design of Electronic Equipment*,
DOI 10.1007/978-3-319-09111-2

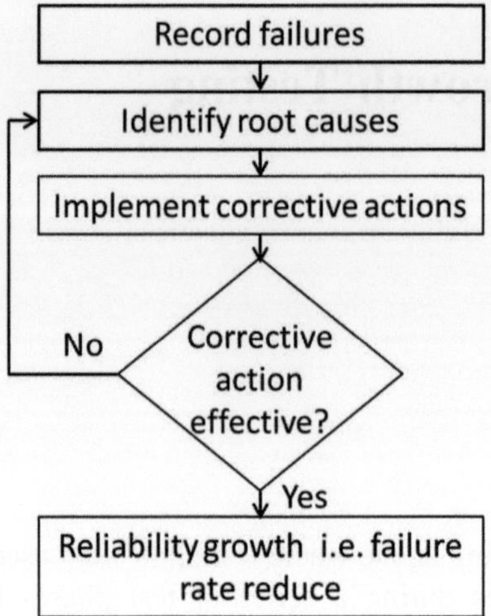

Fig. B.1 Reliability growth process

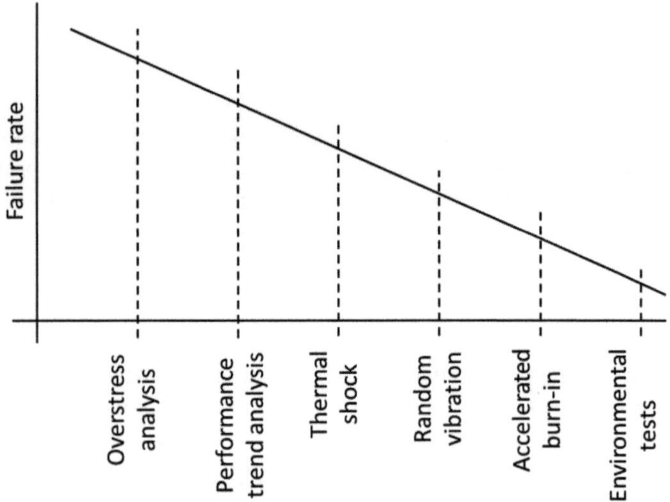

Fig. B.2 Benefit of reliability growth testing

observed in the tests, the benefit of the reliability growth testing is reducing failure rate and it is displayed in Fig. B.2. The failure rate does not indicate quantified values.

Crow (AMSAA) model could be used to estimate individual effectiveness factor for the reliability growth tests by analyzing the available failure data in the tests [1].

Reference

1. Gibson, G.J., Crow, L.H.: Reliability Fix Effectiveness Factor Estimation. In: IEEE Proceedings of the Annual Reliability and Maintainability Symposium. pp. 171–177 (1989)

Appendix C
G_{rms} for Random Vibration Profiles

C.1 About G_{rms}

Random vibration profile relates Acceleration Spectral Density (ASD) in g^2/Hz and frequency in Hz. The random vibration profile shown in Chap. 6 is reproduced below as Fig. C.1. The profile has two curves separated by the crossover frequency, 40 Hz. The square root of the area of each curve is represented by G_{rms}. The G_{rms} value for the entire vibration profile is calculated from the individual G_{rms} values of the curves. Calculation of G_{rms} is explained for the vibration profile, shown in Fig. C.1. The procedure for the calculation is based on the NASA document [1].

C.2 Calculation of G_{rms}: (10–40) Hz

Step-1:

Calculate the number of octaves (O_n) in the frequency band, (10–40) Hz

$$O_n = \frac{\log\left(\frac{F_H}{F_L}\right)}{\log(2)}$$

$$F_H = 40 \text{ Hz} \qquad F_L = 10 \text{ Hz}$$

$$O_n = \frac{\log\left(\frac{40}{10}\right)}{\log(2)} = 2$$

Step-2:

Calculate the dB value.

$$dB = 10 \log\left(\frac{\mathrm{ASD}_H}{\mathrm{ASD}_L}\right)$$

$$\mathrm{ASD}_H \text{ at } 40\,\text{Hz} = 0.015 g^2/\text{Hz} \qquad \mathrm{ASD}_L \text{ at } 10\,\text{Hz} = 0.015 g^2/\text{Hz}$$

© Springer International Publishing Switzerland 2015
D. Natarajan, *Reliable Design of Electronic Equipment*,
DOI 10.1007/978-3-319-09111-2

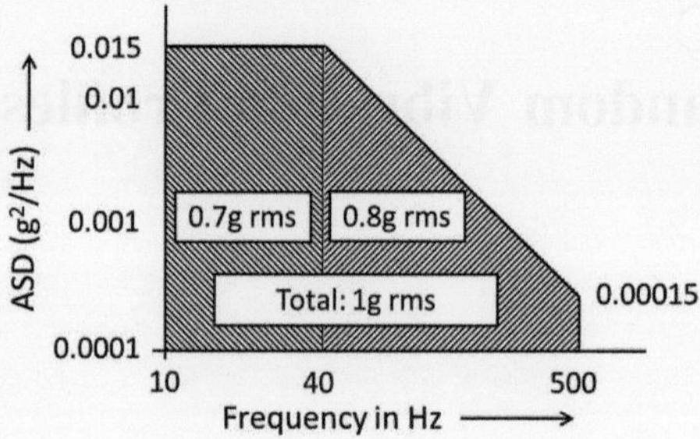

Fig. C.1 Example of random vibration profile

$$dB = 10 \log\left(\frac{0.015}{0.015}\right) = 0$$

Step-3:

Calculate the slope, m.

$$m = \frac{dB}{O_n} = \frac{0}{2} = 0$$

Step-4:

Calculate the area (A) under the curve between the frequencies F_L and F_H. If $m \neq -10\log(2)$, use the following expression to calculate the area, A:

$$A = 10\log(2)\frac{\mathrm{ASD}_H}{10\log(2) + m}\left[F_H - (F_L)\left\{\frac{F_L}{F_H}\right\}^{\left(\frac{m}{10\log(2)}\right)}\right]$$

If $m = -10\log(2)$, use the following expression to calculate the area, A:

$$A = \mathrm{ASD}_L * F_L * \ln\left(\frac{F_H}{F_L}\right)$$

As m $\neq -10\log(2)$,

$$A = 10\log(2)\frac{0.015}{10\log(2) + 0}\left[40 - (10)\left\{\frac{10}{40}\right\}^{\left(\frac{0}{10\log(2)}\right)}\right] = 0.45$$

Step-5:

Calculate G_{rms}.

$$G_{rms} = \sqrt{A} = \sqrt{0.45} = 0.7$$

C.3 Calculation of G_{rms}: (40–500) Hz

$$F_H = 500 \text{ Hz} \qquad F_L = 40 \text{ Hz}$$

$$\text{ASD}_H \text{ at } 500\text{Hz} = 0.00015 \text{g}^2/\text{Hz} \qquad \text{ASD}_L \text{ at } 40\text{Hz} = 0.015 \text{g}^2/\text{Hz}$$

The number of octaves (O_n) in the frequency band, (40–500) Hz,

$$O_n = \frac{\log\left(\frac{500}{40}\right)}{\log(2)} = 3.644$$

$$dB = 10\log\left(\frac{0.00015}{0.015}\right) = -20$$

$$m = \frac{-20}{3.644} = -5.489$$

As $m \neq -10\log(2)$,

$$A = 10\log(2)\frac{0.00015}{10\log(2) - 5.489}\left[500 - (40)\left\{\frac{40}{500}\right\}^{\left(\frac{-5.489}{10\log(2)}\right)}\right] = 0.638$$

$$G_{rms} = \sqrt{0.638} = 0.8$$

C.4 Calculation of Total G_{rms}: (10–500) Hz

$$\text{Total} G_{rms} = \sqrt{A_{(10-40)\,Hz} + A_{(40-500)\,Hz}}$$

$$\text{Total} G_{rms} = \sqrt{0.45 + 0.638} = 1.0$$

Reference

1. FEMCI (Finite Element Modeling Continuous Improvement) Book, GSFC Code 542. NASA, USA (Oct. 2009)

Answers to Quiz and Problems

Chapter-2

1. (i) T (ii) F (iii) T (iv) T (v) F
2. 181 °C. No. It exceeds the de-rating limit by 1 °C
3. 0.4 W

Chapter-3

1. (i) T (ii) T (iii) F (iv) F (v) F

Chapter-4

1. (i) F (ii) T (iii) F (iv) T (v) F (vi) F (vii) F (viii) T

Chapter-5

1. (i) T (ii) T (iii) T (iv) T (v) F (vi) F (vii) F (viii) F (ix) F

Chapter-6

1. (i) T (ii) F (iii) F (iv) T (v) F (vi) F (vii) T (viii) F (ix) F

Chapter-7

1. (i) F (ii) F (iii) T (iv) F (v) F

Chapter-9

1. (i) F (ii) F (iii) F (iv) T

© Springer International Publishing Switzerland 2015
D. Natarajan, *Reliable Design of Electronic Equipment*,
DOI 10.1007/978-3-319-09111-2

Index

© Springer International Publishing Switzerland 2015

D. Natarajan, *Reliable Design of Electronic Equipment,*

DOI 10.1007/978-3-319-09111-2